THE ANALEMMA

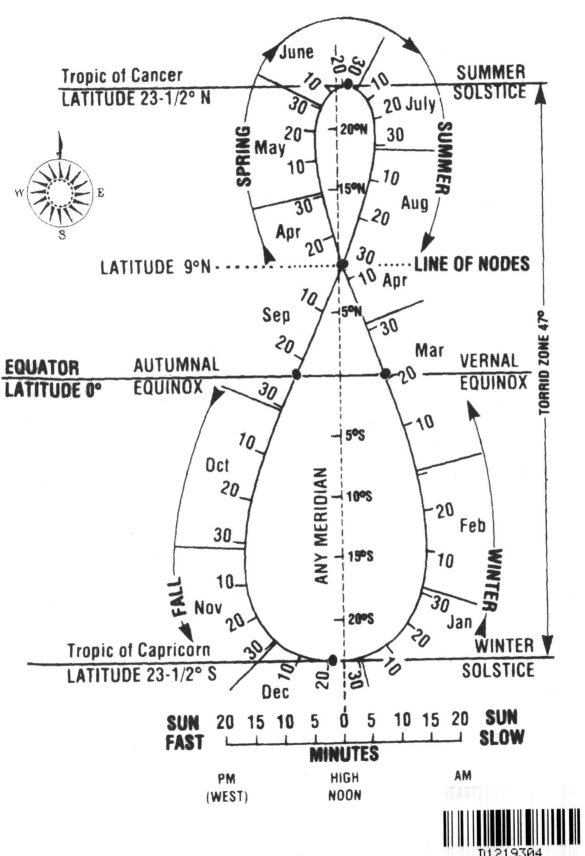

Tropic of Cancer
LATITUDE 23-1/2° N

SUMMER SOLSTICE

June

SPRING

May

Apr

SUMMER

July

Aug

LATITUDE 9°N · · · · · · · · · · · · LINE OF NODES

20°N

15°N

5°N

Sep

Mar

EQUATOR
LATITUDE 0°

AUTUMNAL EQUINOX

VERNAL EQUINOX

TORRID ZONE 47°

Oct

5°S

10°S

Feb

ANY MERIDIAN

15°S

WINTER

FALL

Nov

20°S

Jan

Tropic of Capricorn
LATITUDE 23-1/2° S

WINTER SOLSTICE

Dec

SUN FAST   20 15 10   5   0   5 10 15 20   **SUN SLOW**

MINUTES

PM
(WEST)

HIGH
NOON

AM
(EAST)

W · E · S

CELESTIAL

NAVIGATION

QUICK

In your head calculations of latitude and longitude

FULLY ILLUSTRATED

Over 80 internet websites for further study

Roy T. Maloney

& EASY

DROPZONE PRESS

1709 GOUGH STREET
SAN FRANCISCO, CA 94109

International Standard Book Number: 0-913257-11-7
©Copyright by Roy T. Maloney:2000.

Manufactured in the United States of America by:
    Bertelsmann, Inc., Valencia, CA

Published by:
    Dropzone Press
    Books are available at leading book sellers, or check top of website:
    http ://www .dropzonepress.com:80/great/great1.htm
    e-mail: roymaloney@webtv.net
    URL: http://www.dropzonepress.com
        http://www.themallsf.com

Distributed By:
    Ingram Book Company
    One Ingram Blvd.
    La Vergne, TN 37086-1986
    1-615-793-5000
    Ingram Book Group: http://www.ingrambook.com/
    Ingram Book in Chino, CA
    1-909-590-0680

Library of Congress Cataloging in Publication Data:

Maloney, Roy T.
CELESTIAL NAVIGATION Quick and Easy
In Your Head Calculations of Latitude and Longitude
ISBN: 0-913257-11-7

Cover design and all art: Spiros Bairaktaris, Mountain View, CA   1-650-988-9882

PRINTINGS:
FIRST: April, 2000

*As the midget standing
on the giant's shoulder
can see farther...*

*I dedicate this book to
all the giants who have
helped me.*

The Author

Although it looks like the intrepid sailor on the front cover will meet either a watery grave or be dinner for the shark, he actually got a brief glimpse of the sun and dead reckoned, with his wristwatch, his latitude and longitude. He knew that sailors have been saved by knowing that 20^0 north latitude was the middle of the Hawaiian Island chain. He landed safely on Maui.

CONTENTS

"Man's mind, once stretched by a new idea,
never regains its original dimensions."
Oliver Wendell Holmes, Jr.

INTRODUCTION

It is believed the concepts, to be presented, will stretch your mind.

It takes all of our knowledge to make things simple.

The purpose of this book is to present a navigational system that can help determine your latitude and longitude, or geographic position (GP). With practice you can determine your GP within 1 or 2 degrees.

You need only your brain and a watch.

To obtain a more accurate GP, you will need to study celestial navigation, push a button on a Global Positioning System (GPS) or LORAN, a long range navigation system using time displacement signals.

Advanced celestial navigation systems are more accurate, but it does not follow that you will have a greater understanding.

If the words are not clear it is the fault of the writer. It means that the explanations were not presented in a clear and understandable manner.

We hope these pages will give you an understanding of dead reckoning and navigation.

The author.

PARALLELS OF LATITUDE

Latitude is represented by an infinite number of parallel circles going around the earth from east to west.

They never cross and are all parallel to the equator.

These circles are measured in degrees, north or south of the equator.

The equator is at zero degrees (0°). Note that these are degrees of angle and not temperature.

The equator is a great circle, equidistant from the North Pole and the South Pole

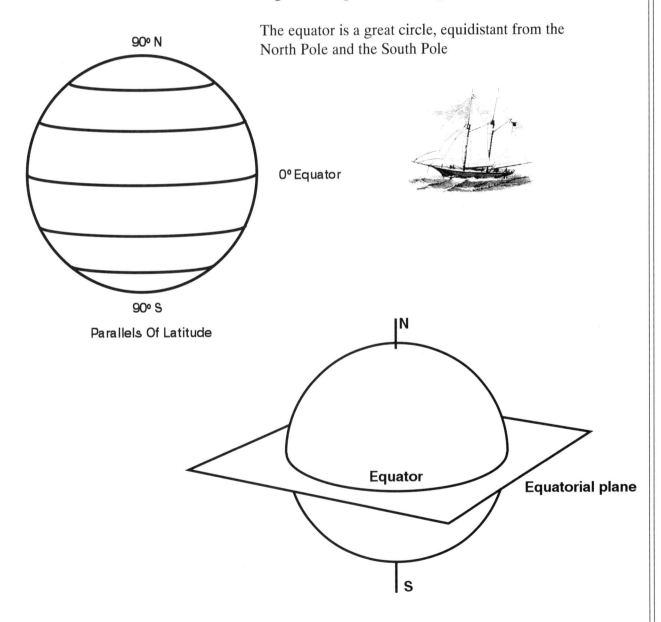

90° N

0° Equator

90° S

Parallels Of Latitude

N

Equator

Equatorial plane

S

MERIDIANS OF LONGITUDE

Longitude is represented by an infinite number of circles that go around the earth passing through both the north and south poles.

The plane of the longitudinal circles pass through the center of the earth. Therefore all the circles of longitude are great circles.

One half of a longitude circle is a meridian. A meridian goes from the north to the south pole.

Longitude is measured east or west of the Greenwich meridian, which is at zero degrees (0°) longitude.

As latitude is measured, north or south, from zero degrees at the equator... so longitude is measured, east or west, from the Greenwich meridian at zero degrees.

The Greenwich meridian goes directly through the Greenwich observatory in England. There is a brass strip set in concrete to mark the starting point of longitude.

Since the entire circle around earth is 360°, the maximum number of degrees you can go east or west is 180°. East is half a circle and west is the other half.

This same logic that applies to longitude, applies to latitude, except you are going from north to south, from zero degrees at the equator. The maximum latitude is either 90° north, or 90° south. These being the north and south poles.

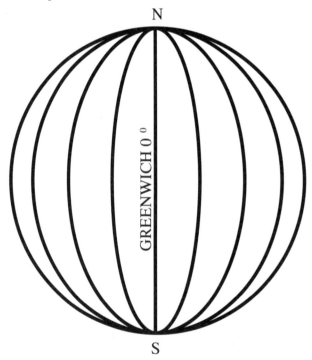

Meridians of Longitude

THEY AIDED CELESTIAL NAVIGATION

Ptolemy
(c A.D. 140)

Marco Polo
(1254 -1324)

Christopher Colombus
(1451-1506)

Nicolas Copernicus
(1473-1543)

Ferdinand Magelan
(1480-1521)

Gerhardus Mercator
(1512-1554)

Sir Francis Drake
(1540-1596)

Tycho Brahe
(1546-1601)

Johannes Kepler
(1571-1630)

Captain James Cook
(1728-1779)

Albert Einstein
(1879-1955)

Neil Armstrong
(1930-   )

The whole theory of the universe is directed unerringly to one single individual - namely to you.
Walt Whitman (1819-1892)

THE NORTH STAR

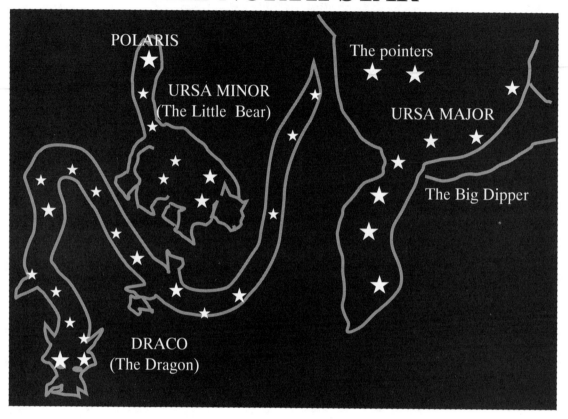

Polaris can be found by using the two pointer stars in the cup of the Big Dipper. Polaris is around five times the distance away from the distance between the two stars.

NORTH STAR

Polaris (the pole star), is the tail star in the Little Dipper (the Little Bear), about 5 times away, from the distance between the pointer stars in the Big Dipper (Ursa Major, the Big Bear).

Polaris has been used for thousands of years to determine true north.

The four stars in the body of the Little Dipper, are a good reference point for magnitude of stars. They are 2nd, 3rd, 4th and 5th magnitude stars.

LUNAR MONTH

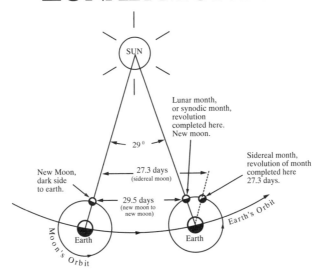

Lunar month, or synodic month, revolution completed here. New moon.

Sidereal month, revolution of month completed here 27.3 days.

29°

27.3 days (sidereal moon)

New Moon, dark side to earth.

29.5 days (new moon to new moon)

Moon's Orbit

Earth's Orbit

Earth

Earth

TIDES

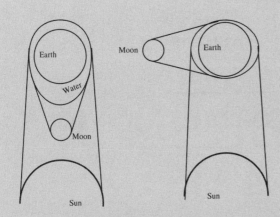

Moon

Earth

Earth

Water

Moon

Sun

Sun

When the sun and moon are aligned, near the full moon and new moon, the tides are highest, due to gravitational pull. This is called "spring tide."

When the sun and moon are at right angles to each other, near the quarter moons, the tidal bulge is lower and called the "neap tide."

Tides are highest when the moon in its elliptical orbit is nearest earth.

When the sun and moon are in line, it causes the tides to rise, about every 12

hours. This is caused by the increased gravitational force.

The moon's gravitational force, is pulling on the water, creating tides, and creates a forced wave to the east.

When the two forces are balanced, i.e., the moon's gravitation and the rotation of the earth, they hold a tidal crest to the east of the moon and not directly under it.

Although the sun is larger than the moon, the sun's tide generating force is only 46% of the moon's force, because of its greater distance.

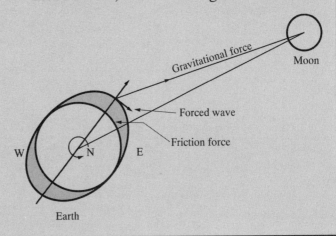

Gravitational force

Moon

Forced wave

Friction force

W

N

E

Earth

CELESTIAL SPHERE
WINTER AND SUMMER

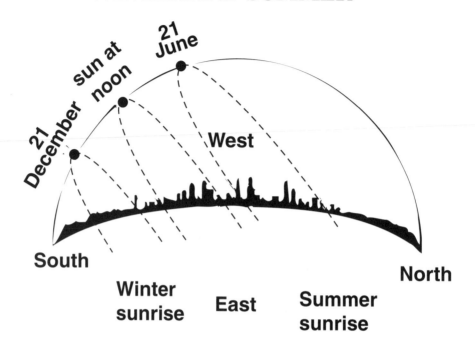

21 June

sun at noon

21 December

West

South

North

Winter sunrise

East

Summer sunrise

SOLAR SYSTEM ON THE MOVE

The sun and our entire solar system, are moving in the direction of the Hercules constellation.

The earth revolves around the sun in an elliptical spiral, never to complete the ellipse.

★ Vega

Sun's path towards Vega

Earth

HISTORY OF THE SEXTANT

It's all done with angles from star to horizon.

1

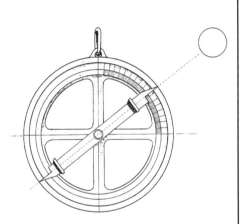

KAMAL

2

QUADRANT

3

ASTROLABE

4

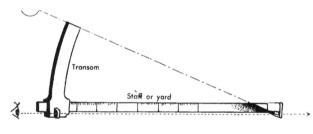

Transom

Staff or yard

BACK STAFF

5

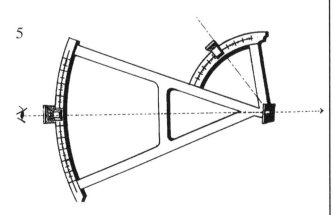

DAVIS QUADRANT

6

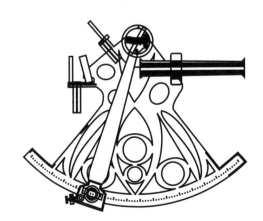

SEXTANT

USING A PROTRACTOR FOR 30⁰

On a 3 foot section of cardboard, or plastic, draw 30°, 15°, and 10° angles.

The sun moves 30^0 in two hours

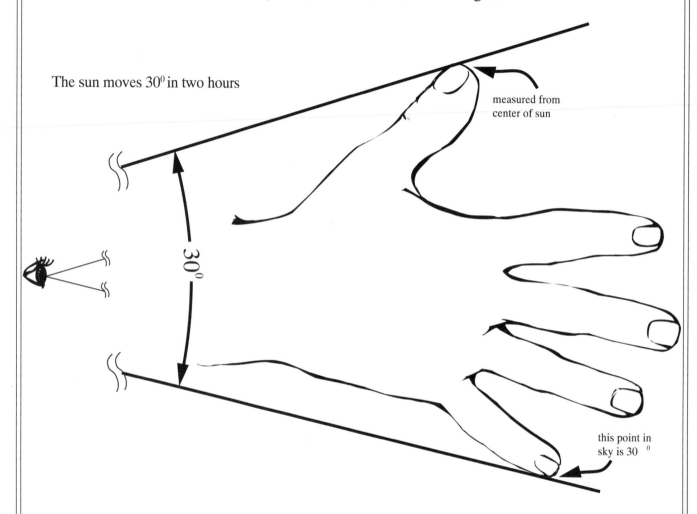

measured from center of sun

30^0

this point in sky is 30 ⁰

MEASURING DEGREES

Place the cardboard angle near your eye and extend your arm, with fingers spread, until it exactly equals 30°. It works with any size hand or arm.

Depending on how far your hand span is from your eye, determines the number of degrees. If you extend your arm in a stiff manner, the hand span can equal 15°. If you extend your arm in a natural way, with your forearm at about a 90° angle, in relation to your upper arm... your hand span will represent 30°.

You must practice this to determine the 30° angle (arc). Once you have mastered the proper position, it will then remain at 30°.

MEASURING MOVEMENT OF THE SUN

Check your extended hand measurement, several times, until you are confident you can duplicate the results.

Do the same with the thumb and fist to define 15°.

Fist alone for 10°, and one finger to equal 2°.

Unless you are still growing, your arm and hand measurements will remain the same. Just as when you extend your arms, your height is the same as the distance between your finger tips.

Human lengths have been used for **thousands** of years. Sailors still measure their **fishing nets** by stretching the net between their **extended** hands. Replicating a fathom of **six feet**.

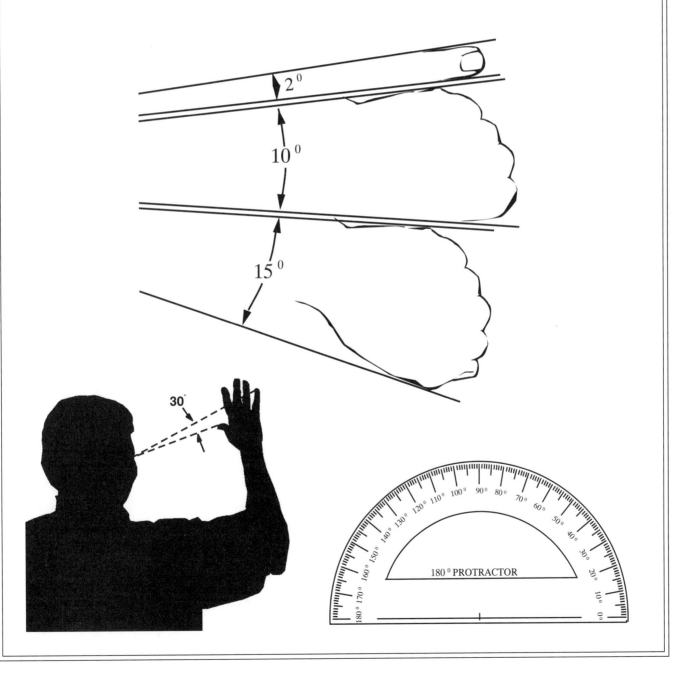

Because stars are so far away, they look like points. Planets look like discs and seem to move against fixed stars.

According to Buddhism there are an infinite number of universes.

"I want to know God's thoughts...the rest are details." Einstein

To visualize infinity ... just close your eyes ... the darkness has no dimensions.

Einstein has shown that mass causes the curvature of space and time. If there were no mass, no stars, no planets and no singularities ... there would be a condition before Genesis.

A pre-Genisis would indicate a condition of Euclidian geometry. It is the only condition that would allow straight lines to infinity.

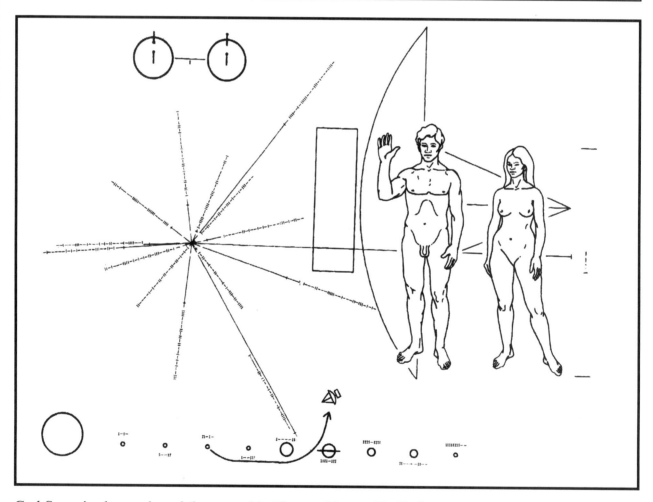

Carl Sagan's plaque aboard the spaceship Pioneer 10, mankind's first attempt to communicate with extra-terrestrial civilizations.

The two circles on top left express the hydrogen atom, just below is its radio frequency.
The circle on the bottom left is our sun. The earth is the third planet from the left.
The height of the man and woman is represented by the size of the spaceship to their left.

LATITUDE AT NIGHT

It is believed the most simple "dead reckoning" involves using the north star (Polaris).

The north star is directly over the north pole (within one degree), its latitude is 90 N°. The north star is at 90° in relation to the plane of the equator.

If you visualize a line from any position on the equator, to the center of the earth, and then vertically to the north pole, and north star...you

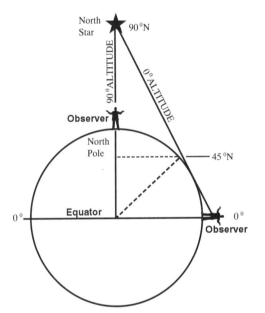

have created a 90° angle. The north star is at zenith to the north pole.

If you observe the north star, from any position at the equator, it is on the horizon at 0° latitude. The equator is at 0° latitude.

As you move south, away from the north star, your latitude changes from 90°N to 0°. The north star goes lower and lower on the horizon.

At the half-way point your latitude is 45°N, the angle to the north star is also about 45°N. The angle, also called altitude, of the north star from your position, using the horizon as your base, is your latitude.

This system has been known at least 5,000 years.

This identical logic applies in the southern hemisphere with the southern cross.

A submariner navigator, on-board a U.S. submarine, under the polar ice cap, at the north pole, entered in their log "Latitude 90°N, Longitude: Infinite."

MEASUREMENT CHECK

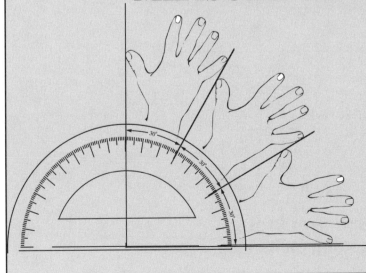

There is a very simple way to see if your open hand measurement is accurate.
When placed before your eyes your hand should represent 30^0. Measuring from a vertical, in the sky to the horizon is 90^0. From a vertical to the horizon should be exactly three hand spans, or 90^0.
When it is afternoon. note the sun at 30° above the horizon. The sun will then set in two hours.

THE SOUTHERN CROSS
USED FOR DETERMINING TRUE SOUTH
The stars above the south pole.

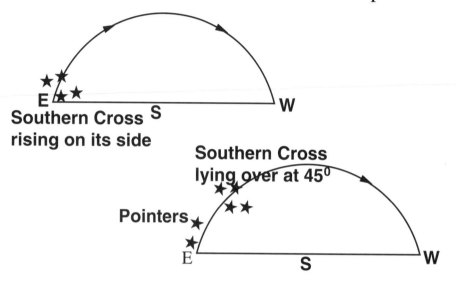

**Southern Cross
rising on its side**

**Southern Cross
lying over at 45⁰**

Pointers

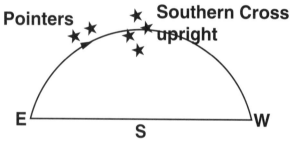

Pointers

**Southern Cross
upright**

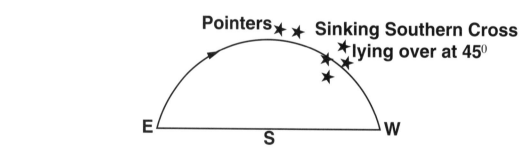

Pointers **Sinking Southern Cross
lying over at 45⁰**

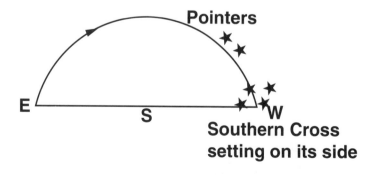

Pointers

**Southern Cross
setting on its side**

THE ANALEMMA

(pronounced: an-a-lem-a)

The word analemma, originally derived from the Greek word, for a pedestal of a sundial.
By observing the date on the analemma, you can determine the latitude of the sun on that day.

Example: 21 June, 23 1/2°N, the summer solstice.

21 December, 23 1/2°S, the winter solstice.

Although the dates of the solstices may vary, the above serves us well for dead reckoning.

When you have determined the latitude of the sun, as given on the analemma, for a specific date,
you then need to know your horizontal angle, from the sun in degrees.

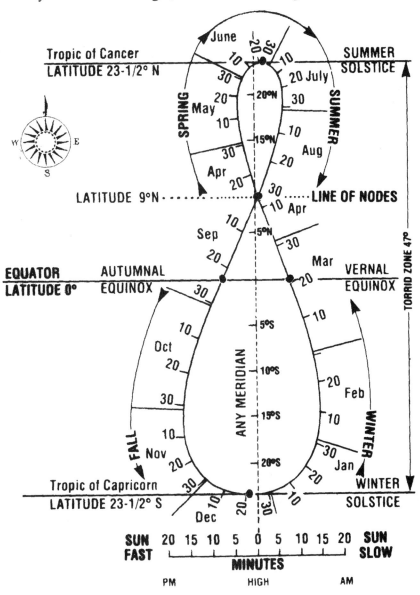

CONSTRUCTING AN ANALEMMA
IN YOUR HEAD USING 7 POINTS

Constructing an analemma on a blank piece of paper is quite easy. You can construct the outline with just seven sun positions. These 7 points of the sun are as follows:

1. The number 21 is a key date to remember and serves our purpose of dead reckoning very well, e.g. March 21 is vernal equinox.

2. Draw a vertical line for the meridian. Place on the top, or north, of any one meridian 21 June. This is on the Tropic of Cancer and is at 23 1/2° north latitude.

3. Place on the bottom, or south, 21 December. This is the Tropic of Capricorn, at 23 1/2° south latitude.

4. Draw in the equator, half way between the two Tropics. This is 0° latitude.

5. The only place the analemma crosses is at 9° north latitude.

6. Just below half way between the equator and the Tropic of Capricorn, is 15° south latitude. At this location go east and west about an inch, place 2 points, to represent 15 minutes of time, that the sun is fast or slow.

7. On the equator, to the east and west of the meridian, place two points at 7.5 minutes, or half of the 15 minute mark already indicated in number 6. To the east the point

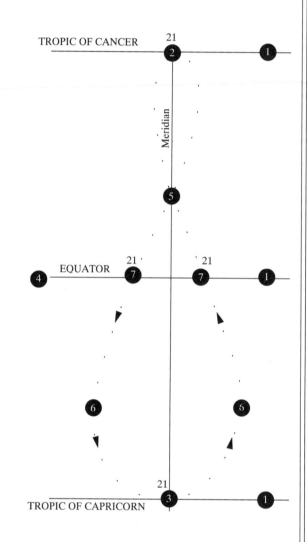

is at 21 March, the vernal equinox. To the west the point is 21 September, the autumnal equinox.

From the above seven points, you can connect the dots to form an analemma. Note that you have created the four seasons and **you can estimate the path of the sun for the entire year**.

Try it a few times and it becomes very simple.

INTERESTING FACTS ABOUT THE ANALEMMA

The analemma defines the solstices, between the Tropic of Cancer and the Tropic of Capricorn...latitude 23 1/2°N and latitude 23 1/2°S.

The sun at zenith (90° altitude), does not go north of latitude 23 1/2°N, or south of latitude 23 1/2°S.

The total movement of the sun, as indicated on the analemma, is 47°, north and south, and defines the Tropics. This represents all of the latitudes between the Tropic of Cancer and the Tropic of Capricorn. When we speak of the movement of the sun, it represents the virtual sun.

Johannes Kepler (1571-1630)

The earth tilts on its axis 23 1/2°, in relation to the sun.

The Tropic of Cancer is 23 1/2° north of the equator.
The Tropic of Capricorn is 23 1/2° south of the equator.
The arctic circle is 23 1/2° from the north pole.
The antarctic circle is 23 1/2° from the south pole.
This represents **five** 23 1/2° angles.

Because the path of the earth around the sun is an ellipse and not a circle, the analemma is only on its meridian at exactly 12 noon, only 4 times per year. The 4 times are: the 2 solstices and the crossing at 9°N two times per year. (Note the analemma).

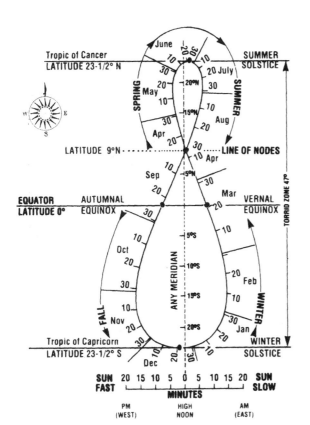

Other than the 4 times per year, the sun is either east, or before (slow), the analemma meridian, or west of the meridian, or past (fast).

The sun at 12 noon is either slow or fast, in relation to its meridian, by a maximum of about 16 minutes of time.

This difference in the position of the sun, in relation to its meridian, is known as the "Equation of Time."

The term solstice is very descriptive, it means sun stop.

When you look at the sun, with sunglasses, shining on your location and half of the earth, it is not obvious that the sun, at 12 noon, is only at zenith (90°), or directly overhead at only one spot on earth, or one specific latitude.

If you are standing in Shanghai, Los Angeles, Chicago, London, Madrid, Paris, Berlin, or Moscow...you will always look south to view the sun. The sun will never be over your head.

If you are standing in Sydney, Auckland, Buenos Aires, or Capetown...you will always look north to view the sun. The sun will never be over your head.

Galileo Galilei (1564 - 1642)

If you are standing in Honolulu, Mexico City, Barbados, Rio de Janero, Nairobi, Bombay, Bangkok, or Manila...there will be two days per year when the sun will be directly over your head.

The sun spirals 24 hours per day, from the Tropic of Cancer to the Tropic of Capricorn. The analemma uses just one point of time per day, namely 12 noon. The path of the sun, is in reality the orbit of the earth around the sun, between the two Tropics. This orbit of earth, is the ecliptic.

A simple way to think of the analemma, is to imagine you took a photo of the sun, on one negative, ever day for one year, at 12 noon, in front of your house. When you developed this time-lapse negative you would have a figure 8...an analemma.

Another way to visualize the analemma is to start at the Tropic of Cancer, at any single meridian, at the summer solstice. Then proceed at 12 noon to move so that the sun hits you directly on the top of your head, so there is no shadow. If you do this for one year, you will return to the Tropic of Cancer. You will have traversed a figure eight path...an analemma.

By observing the date on the analemma, you can determine the latitude of the sun on that day.

Nicholas Copernicus (1473-1543)

FIX

Navigators express their position, first in latitude and then in longitude. This is the same system used when expressing the position on a grid. First is (X), the horizontal line and then (Y), the vertical line.

For example: 36°N, 130°W, means 36°N latitude and 130°W longitude.

Every point on earth can be expressed in latitude and longitude, in degrees, minutes and seconds.

There are 360° in a circle, 60 minutes in a degree and 60 seconds in a minute. These terms relate to measurement of an angle, or arc, not time.

For purposes of clarity and dead reckoning, we will limit the system to degrees.

Latitude is like the brim of a hat, longitude is like the crown.

It is confusing for many people, to remember the difference between latitude and longitude. The reason may be because, latitude is horizontal, but measured vertically. Longitude is vertical, but measured horizontally.

If you see a number in degrees, that states north or south, it refers to latitude.

If you see a number in degrees, that states east or west, it refers to longitude.

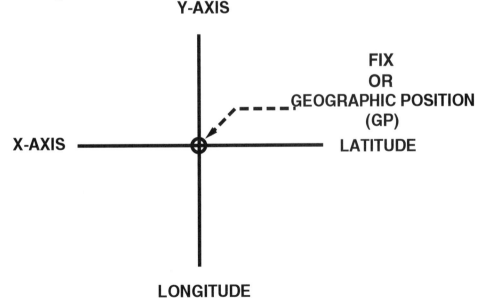

To describe where you are, use latitude and longitude

WHERE IN THE WORLD ARE WE?

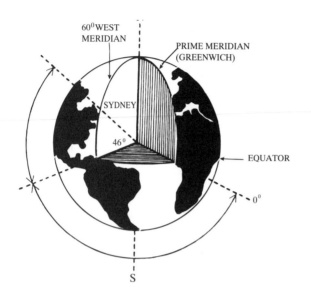

Let's say you are standing in the middle of Sydney, Canada and wanted to describe where you are in terms of world location. You could start by locating the Prime Meridian that runs through Greenwich, England. Meridians are the great circles running north and south through the center of the earth. Where the Prime Meridian intersects the equator is the starting point, or 0° for measuring east or west around the equator. Ancient man spoke of 'medius dies,' or middle of the day, when the sun was directly overhead. Medius dies eventually became meridian. If you go in an arc in a westerly direction 60° from the Prime Meridian you arrive at the 60° West meridian at the point it crosses the equator. If you now proceed due north on this meridian for 46° you will arrive at Sydney, Canada. Its location on earth is 46° N, 60°W.

COMPASS ROSE

Directions and descriptions by points of the compass are more specific. To be more accurate, directions can be given in exact degrees, from 0° to 360°, North is 0° or 360°, East is 90°, South is 180° and West is 270°.

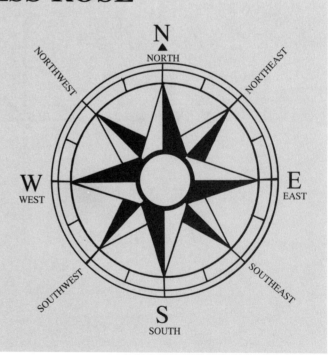

DETERMINING LATITUDE

If you know the latitude of the sun, and you know the number of degrees you are from the sun, either north or south, you then know your latitude. On any given day, the latitude of the sun, is easily determined with the analemma.

The analemma is a serries of points, plotted on the surface of the earth, that when connected, trace out a figure eight (8). Each point represents the sun at high noon. Each point represents one day and all of the points represent one year.

Each day represents a slightly different declination (angle), or latitude. The full range of the sun's one year movement, is from the Tropic of Cancer to the Tropic of Capricorn.

The analemma represents, the sun, each day at high noon, or zenith, for just one meridian. It can be the meridian of your choice.

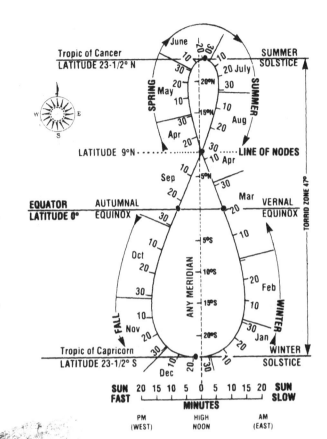

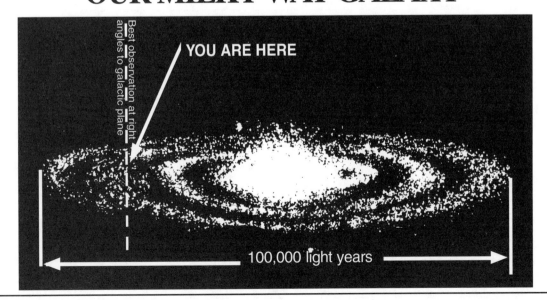

OUR MILKY WAY GALAXY

YOU ARE HERE

Best observation at right angles to galactic plane

100,000 light years

*"Difficulties of human behavior are more
difficult than the difficulties of technology."*
Dr. Edward Teller

NOTES ON LATITUDE

Standing due west, or east, measure with your hand the number of degrees (horizontaly) you are from the sun.

South can be determined with a compass, or at 12 noon, you will have a true north and south line. (It will be very exact if you correct for the "equation of time").

To explain in detail the difference between true north and magnetic north is the subject of another book. For our purposes of dead reckoning, the zenith, or highest point of the sun is very reliable.

A compass (magnetic) can be checked against the north star to determine true north, as Columbus did.

The sun (virtual) travels from east to west in an arc. The top of the arc is the zenith point. This zenith point can be estimated to establish a north/south line. Any upright will cast a shadow in an arc. This is the principle of the sun dial.

It is a virtual sun because the sun does not go around the earth, but it appears to do so, because of the earth's rotation.

When you face due south and extend your arms, your right hand is facing west and your left hand is facing east. You have just established the cardinal points of the compass: north, south, east and west.

When facing west, or east, visualize a vertical line, a 90° angle, going up to the height of the sun. Then hand measure the number of horizontal degrees, from the center of the sun, to the imaginary point in the sky, on the vertical line, at the same height as the sun.

This is the number of degrees you are from the sun.

MEASURING LATITUDE

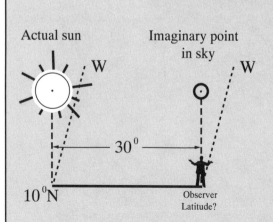

Actual sun

Imaginary point
in sky

W

W

30°

10°N

Observer
Latitude?

This is a new concept, for you are measuring degrees, left or right, from an imaginary vertical line from the sun and an imaginary vertical line from your position... extending to the same height as the sun.

In the diagram the sun is at 10° North, as indicated by the analemma.
If you are 30° north of the sun, you are therefore at 40° North latitude.

Note you do not need a horizon.

KNOW WHERE YOU ARE

From the analemma we know the sun for this day is at 20^0 N. We measure from the middle of the sun to a vertical point in the sky.

If we measure 17.5^0, we know we are at latitude 37.5^0 N. We measure 14^0, we know we are at latitude 34^0 N.

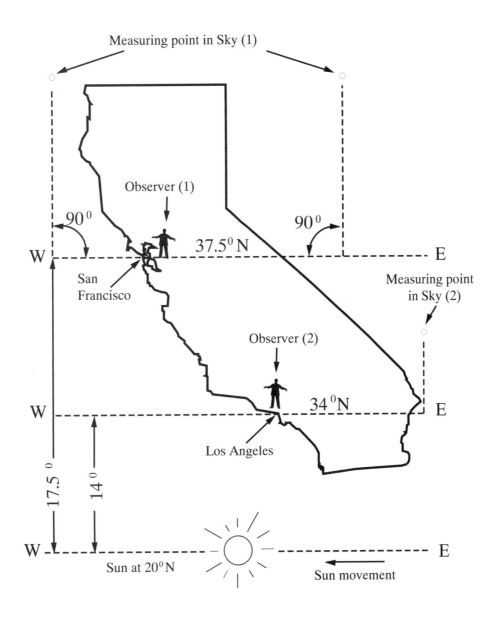

WHERE AM I ON THE EAST COAST?

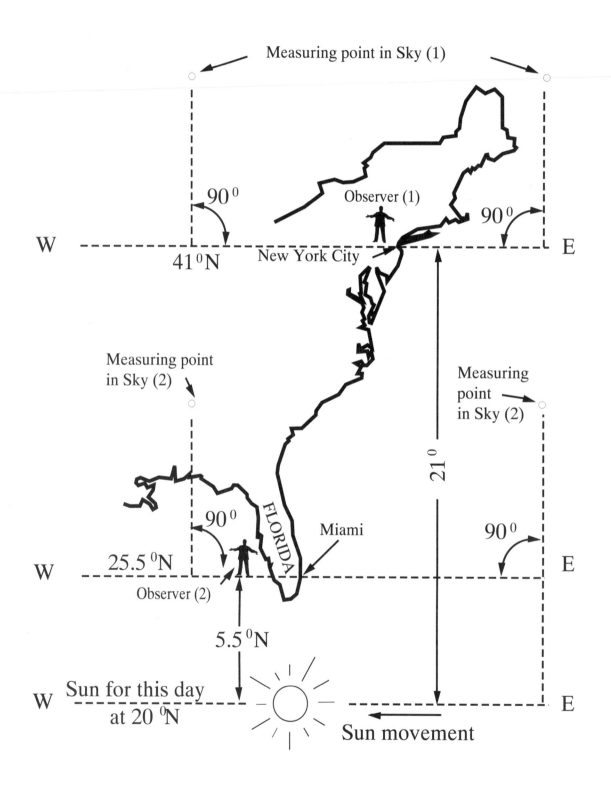

Measuring point in Sky (1)

90°

Observer (1)

90°

W - E
41°N   New York City

Measuring point
in Sky (2)

Measuring
point
in Sky (2)

21°

FLORIDA

Miami

90°

90°

W - E
25.5°N

Observer (2)

5.5°N

W   Sun for this day   - - - - - - - - - - - - - - - - E
    at 20°N

Sun movement

Frigate under full sail.

ZENITH

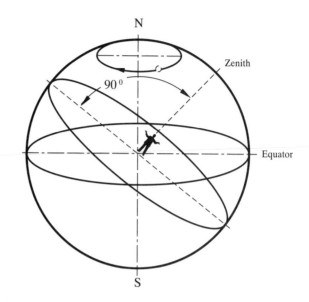

Many people speak of the "direct rays of the sun". The more accurate word is zenith. Zenith is an important concept. When the sun hits you on top of your head and there is no shadow, the sun is at zenith.

When the sun, or a star is at zenith, it is vertical to you as an observer. When a celestial body is vertical to you, or to a point on earth, it is at zenith.

When a celestial body is at a 90° angle, between the body and the horizon, it is at zenith. So you can speak of zenith as vertical, 90° angle and directly overhead.

At any instant of time, every celestial body is at zenith (directly above), some point, or geographic position (GP) on earth.

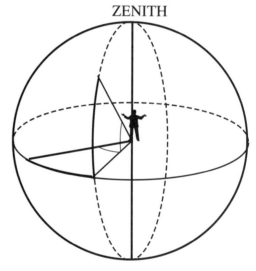

The altitude of a celestial body, is the angle between the body and the earth's horizon.

If you have a point on earth, each day at high noon, for one year, for all of the meridians... you have created an ecliptic.

You have created a circle around the earth, that represents the path of the virtual sun. It is a virtual sun, because the sun, in reality, does not go around the earth.

As we know, the earth revolves around the sun. The ecliptic remains the same. It is the path of the earth around the sun.

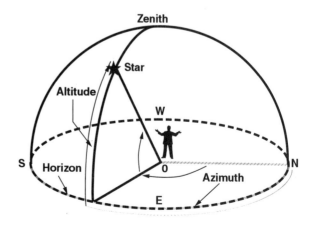

LA-TIT-DUDE AND LONG-DUDE

Latitude and longitude are confusing to most people.
Think of *latitude* (la-tit-dude) as the measurements on a woman,
and *longitude* as a tall basketball player (long-dude).

Latitude from the Latin is ***latus***, wide.
Longitude from the Latin is ***longus***, long.

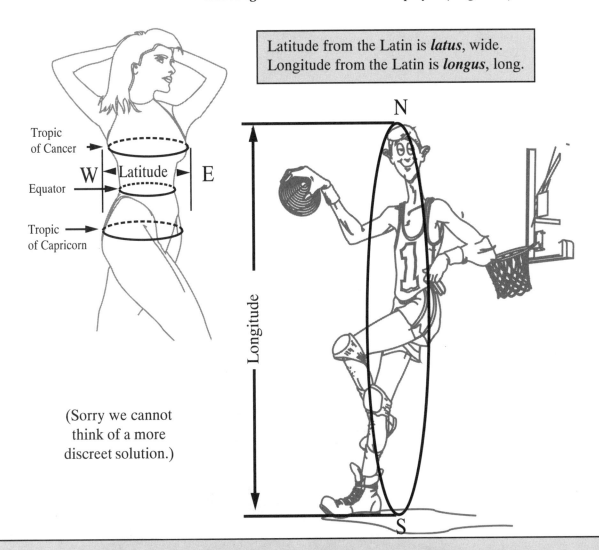

Tropic of Cancer

Equator

Tropic of Capricorn

W ◄ Latitude ► E

N

Longitude

S

(Sorry we cannot think of a more discreet solution.)

Dr. Edwin Hubble exposed the Andromeda galaxy for over 30 hours and for the first time exposed individual distant stars. He also showed that the galaxies were moving away, the "red shift". The universe is expanding.

A super nova can produce more energy, than our sun in its lifetime.

Einstein once said, "The human mind is not capable of grasping the universe."

Stephen Hawking has said, "The prospect of an early death concentrated my mind wonderfully." "The universe is a giant casino with dice being rolled or a wheel being spun at every occasion." "When we incorporate the uncertainty principle into Einstein's general theory of relativity, we're close to understanding the universe."

The *Thames*, East Indiaman, 1424 tons, built for, and employed by, the Honourable East India Company. James Keith Forbes, commander.

DETERMINING LONGITUDE

(usually by radio signal or phone)

If you set your watch at Greenwich time and your watch reads 8pm, or 2000 hours, and the sun at your location is at zenith, or 12 noon, you know you are 8 time zones from Greenwich.

Greenwich time is also called Zulu time, in the military, as Zulu stands for Z, or zero meridian.

Since the sun travels from east to west, in the sky, it is 8 hours later to the east, or 8pm at the Greenwich meridian.

Your location is therefore, at the center of a time zone, to the west. Eight time zones times 15° is equal to longitude 120°W.

If you understand the above, you understand the concept of determining longitude.

If you were at the equator at longitude 120°, you would be 120 times 60nm, or 7,200nm from the Greenwich meridian. That is how you calculate distance from the Prime Meridian.

Since 15° is equal to one hour. Each degree is equal to 4 minutes of time, 60 divided by 15. So if your watch (chronometer), on Greenwich time, indicated 8:04pm (2004 hours), with the sun at the 12 noon position at your location, you would be located 4 minutes to the west of the center of the time zone. Your location would be longitude 121°W.
The width of the analemma represents time.

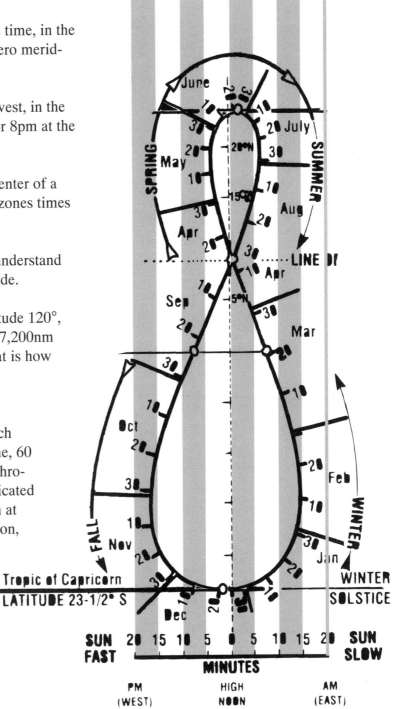

LONGITUDE AND TIME ZONES

All of the infinite circles of longitude are great circles.

The planes of longitude all go through the center of the earth, the north and south poles, and extend to infinity. They are not parallel.

Longitude is based on time from the Greenwich meridian, the Prime meridian. Greenwich is at zero degrees. Longitude is measured east or west, in degrees, from 0°. As the equator, at 0° is to latitude, Greenwich at 0° is to longitude.

The maximum distance in degrees, either east or west, is 180°, or 1/2 a circle. East 180° and west 180° is equal to a full circle of 360°. The opposite meridian from Greenwich is the 180° meridian, which is the International Date Line.

Since there are 360° in a circle and 24 hours in a day, when you divide 360 by 24, their are 15° in each hour. This is the movement of the sun, measured at the equator.

The earth actually moves around the sun, but the sun appears to move around the earth, hence it is called a virtual sun.

Since the sun moves 15° in one hour and there are 24 hours in a day, we call each 15° a time zone. There are 24 time zone going around the earth.

The middle of each 15° time zone, is the center of that time zone. So, for example, the Greenwich meridian is the center of a time zone and has 7.5° to the west and 7.5° to he east.

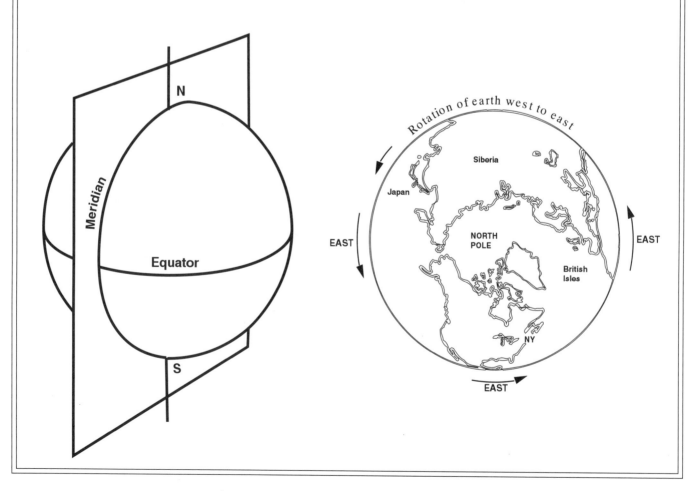

LONGITUDE BY TIME

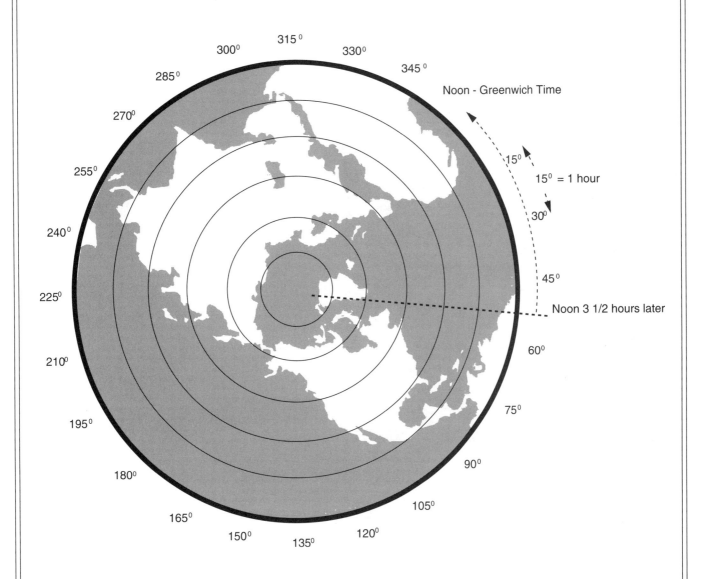

- 315°
- 300°
- 330°
- 285°
- 345°
- 270°
- 255°
- 240°
- 225°
- 210°
- 195°
- 180°
- 165°
- 150°
- 135°
- 120°
- 105°
- 90°
- 75°
- 60°
- 45°
- 30°
- 15°

Noon - Greenwich Time

15° = 1 hour

Noon 3 1/2 hours later

Note 24 time zones for 24 hours (one day).
15 degrees is equal to one hour.

JOHN HARRISON

John Harrison (1767) the man who found longitude.

John Harrison's watch "*Old number Four*", used to find longitude. More accurate time, more accurate longitude.

DEGREES OF ARC (minutes and seconds)	TIME MOVEMENT OF SUN (minutes and seconds)	DISTANCE ON EQUATOR OR GREAT CIRCLE
360^0	24 hrs	21,600 NM circumference of earth
15^0	1 hour	900 NM
1^0	4 minutes	60NM
1'	4 seconds	1NM 1.15 statute miles

The "EQUATION OF TIME"

You probably were aware that there is a difference between sun time and clock time. Not as well known is that this difference in time is represented by the width of the analemma.

Consider clock time as exactly on its analemma meridian at 12 noon. When it is not on clock time the sun is either before the meridian (ante meridian), or past the meridian (post meridian). The sun is either fast or slow, when it is not exactly on its meridian at 12 noon. This difference between time on your clock and sun time is the "equation of time."

When the sun is to the west of its meridian at 12 noon, the sun is fast. When the sun is to the east of its meridian at 12 noon, the sun is slow.

Four times per year the sun is exactly on its meridian at 12 noon. Sun time and your clock are exactly the same. The four times are: at the two solstices and at the two times the analemma crosses its meridian at latitude 9°N. At these 4 times the sun is at zenith and your clock says 12 noon. For the rest of the year the sun is either fast or slow.

For example, the widest part of the analemma is at latitude 15°S, at this time the sun is off of clock time by about 16 minutes. The two days on the analemma with the greatest difference from clock time are on about, 10 February (slow) and 1 November (fast).

If the sun is 16 minutes slow, then your clock will say 12:16pm, when the sun is on its meridian and at zenith (90° altitude).

If the sun is 16 minutes fast, your clock will indicate 11:44am when the sun is at zenith on its meridian. The equation of time is a result of the earth's revolving around the sun in an ellipse and not a circle.

You can think of vertical lines, the same as meridian lines, representing time, from the meridian at zero to the east and west, a maximum of 16 minutes.

From personal experience...it is easier to remember the shape and width (time) of the analemma, than to try and remember, a grid for the year, representing days and time difference. A diagram is easier to remember, than a bunch of numbers.

The analog (diagram) is easier to remember than the digital.

KEPLER EQUATIONS

"The form of the orbit of any planet around the sun is an ellipse with the sun as one of the foci." Keppler's first law

The planet, at times, must go faster to create an equal area. The sun creates one focus, the mass of the universe creates the second focus.

Keppler tells us that in his "law of areas" that the line connecting a planet to the sun sweeps out equal areas in equal periods of time. Astronauts used his equations to get to the moon.
The dark areas 1 and 2 are equal in area. The earth moving faster when near the sun.

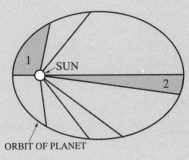

ORBIT OF PLANET

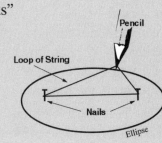

Ellipse

TIME ZONES

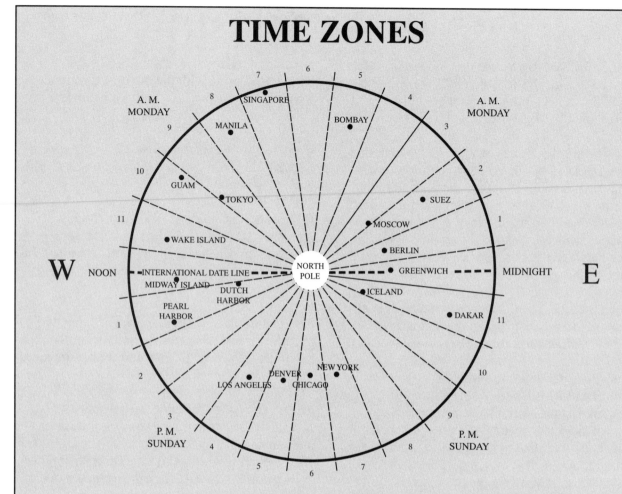

Note that both the Greenwich meridian and the International Date Line are in the middle of their time zones. When it is 12 noon at the IDL it is 12 midnight at Greenwich.

The IDL and Greenwich represent 2 time zones. There are 11 time zones to the east and 11 to the west. for a total of 24. This represents 24 hours and one day. To be all the same day, of the week. it would start and end at 12 midnight at the IDL

When the sun is directly over the center of a time zone, it is the same time in that zone, 7.5° east and 7.5° west, or the entire time zone. The time zone goes from the north pole, to the south pole.

When the sun is directly over the Greenwich meridian at 12 noon it is 11 am (ante meridian) at the time zone to the west and 1pm (post meridian) in the time zone to the east.
In one hour the sun will have moved 15° from the center of one time zone, to the center of the next time zone.

Every hour it will be 12 noon at the center of a time zone, as the sun moves from east to west. After the sun moves through all the time zones, it returns once again to 12 noon at the Greenwich meridian.

Note that when it is 12 noon at the Greenwich meridian, it is 12 midnight at the exact opposite meridian of 180°. It is the start of another day at the International Date Line.

TIME AND DISTANCE

Using round numbers, we know the circumference of the earth is 21,600nm (nautical miles). A nautical mile is 1.15 statute miles. One statute mile is 5,280 feet.

Dividing 21,600 by 24 time zones is equal to 900nm, for each time zone. A time zone is 15°, or 900nm. Each degree is therefore 60nm at the equator, 900 divided by 15. One degree is equal to 60nm, at the equator, on any great circle of longitude, or on any great circle on the globe, it is a very handy reference point.

A great circle, is a circle on a spherical surface, with the plane of the circle passing through the center of the sphere.

Once you know that 60nm is one degree at the equator, you can calculate the cicumference of the earth in your head. You just multiply 60 times 360, the answer being 21,600nm.

One degree at the equator is 60nm and as you go north or south, one degree becomes smaller, when you measure east and west.

At the poles a degree is a point, or zero nautical miles. This is because the circumference of the earth is becoming smaller, as you move away from the equator.

So going away from the equator, measuring east to west, the distance goes from 60nm to zero. A degree of longitude stays at 60nm, because it is a great circle, but a degree of latitude becomes smaller.

You can dead reckon the distance of a degree of latitude, by knowing your longitude. If it is 60nm at the equator and goes to zero nm at the poles...a good guess at the halfway longitude of 45° would be 30nm, or 1/2 of 60nm.

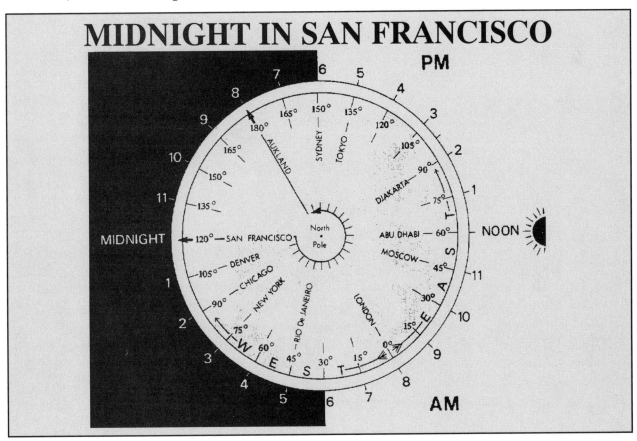

MIDNIGHT IN SAN FRANCISCO

THE EQUATION OF TIME AND LONGITUDE

The equation of time must be considered when determining longitude.

The logic of 4 minutes of time, equal to 1° of longitude does not change. Working with the equation of time is the last step, finally, of dead reckoning longitude.

In our example of the sun being at zenith, at 12 noon at our location and the Greenwich time reading 8pm, 2000 hours military time...you are 8 time zones from Greenwich, or longitude 120°W.

If however, it was one of the 361 days when the sun is not exactly on our meridian at 12 noon...you must add or subtract from the longitude 120°W. For example, if the analemma indicated that the sun was 8 minutes slow, you would know the zenith of the sun was at 12:08pm by clock time.

You would correct for the 8 extra minutes, by subtracting the 8 minutes, or the equal of 2°.

Your longitude by dead reckoning is 118°W and not 120°W.

If on the other hand the analemma indicated the sun was 8 minutes fast, by clock time. The zenith of the sun would be at 11:52am.

You would then add 2°, for a longitude of 122°W and not 120°W.

As with any totally new concept, it takes a little time and patience to do longitude with ease.

SUMMARY

This explanation can be put in summary: observe the zenith of the sun at your location, check the Greenwich time, determine the number of time zones from Greenwich, use the analemma to determine if the sun is fast or slow for that day, and correct for the equation of time.

That's it for longitude and we have previously done latitude, using a system to determine the number of degrees from the equator. Latitude uses angles and longitude uses time.
You can now dead reckon your geographic position on earth.

WHICH WAY IS THE SUN MOVING?

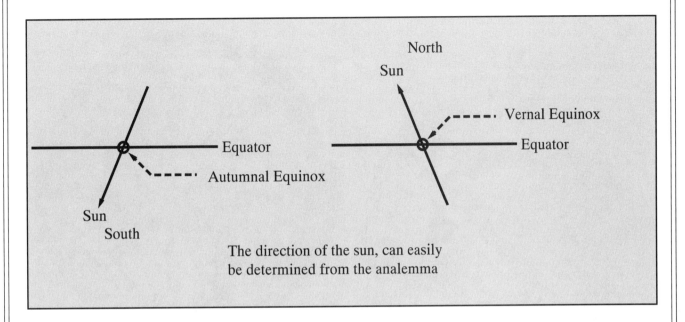

North
Sun

Vernal Equinox

Equator

Equator

Autumnal Equinox

Sun
South

The direction of the sun, can easily
be determined from the analemma

ANGLES BY THE ANCIENTS

Note the angle measurement instruments in the Octagon
Room at the Greenwich Observatory at the time of John
Flamsteed (1646-1719), the First Astronomer Royal. The
men responsible are observing the lunar positions for
Newton's work on the moon.

ARMILLARY SPHERES

Hundreds of years ago, beautiful sculptures were made, depicting the signs of the zodiac on the ecliptic, crossing the celestial equator. The earth was in the center. This sculpture is called an "armillary sphere."

The armillary sphere, correctly has the "celestial Tropic of Cancer" and the celestial "Tropic of Capricorn."

It also correctly has, the "celestial arctic circles" and the "celestial equator."

To bring the armillary sphere up-to-date...the sun should be in the center, with the earth orbiting in the zodiac.

In about a thousand years, the "constellation of Aquarius", will be the vernal equinox.

Your finger with an accuracy of about 2°, is very accurate when standing next to a person. You are within an azimuth, or bearing, of one half inch. At the edge of the known universe, you are accurate within a few trillion miles.

On an ordinary road map, you probably would have no trouble finding A1, or D7, These are grid coordinates, with the alphabet on the left and right, and the numbers on the top and bottom.
This is identical in principle, to giving coordinates in latitude and longitude.

CIRCLE OF EQUAL ALTITUDE

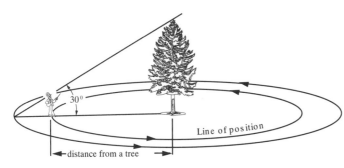

distance from a tree

Line of position

If you were 90 feet from a tree and the angle to the top of the tree was 30⁰...you would have to stay at a 30 degrees angle to continue to look at the top the tree as you walk around it.

This line of position is the same principle used when a sextant gives you an angle to a star and a line of position (LOP).

ANALEMMA SHAPES

All planets and celestial bodies that orbit any sun ... either known or unknown, have an analemma. The tilt and type of orbit determine the shape.

1. A planet with a perfectly circular orbit, meaning no equation of time and therefore no width, but DOES have TILT, it then has only a vertical line as an analemma. It has height but no width.

2. A planet with NO TILT, but an

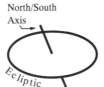

North/South Axis

Ecliptic

ELLIPTICAL or non-circular ORBIT, has a horizontal line on the equator as an analemma. It has the tropics on the equator, no seasons and does have an equation of time. It has width, but no height

3. A planet with NO TILT and a perfect CIRCLE as an orbit. has no seasons, no equation of time. It has no width and no height. The analemma is a point.

FRAME OF REFERENCE

60 MPH

60 MPH

To the observer each car is going 60 MPH. To the driver of the car, the other car driver is going 120 MPH. Your *"frame of reference"* is the key to the theory of relativity. So it is with the stars.

> "Exploration is really the essence of human spirit"
> Frank Borman

OUR PLANETARY SYSTEM

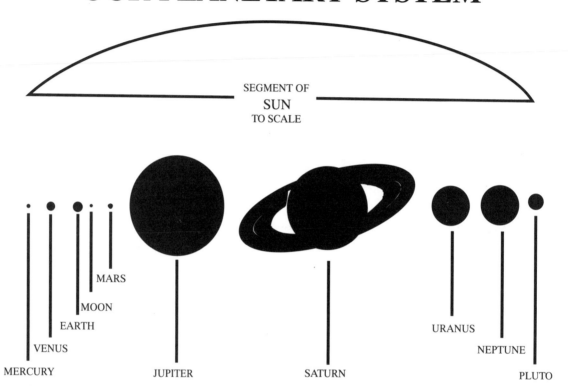

SEGMENT OF
SUN
TO SCALE

MARS

MOON

EARTH

VENUS

MERCURY

JUPITER

SATURN

URANUS

NEPTUNE

PLUTO

To remember the order of the planets try:
Mankind's **V**erdant **E**arth **M**ust **J**ourney (as a) **S**tar **U**nites **N**ine **P**lanets

Mercury-Venus-Earth-Mars-Jupiter-Saturn-Uranus-Neptune-Pluto
Marilyn Vos Savant

FOG INDEX

The most clear and succinct sentence, considered as *"zero fog index*" (degree of clarity), is "sighted sub, sank same." An attempt has been made to achieve that goal, with the prose on celestial navigation. Already this book has taken me, from local obscurity, to national oblivion.

An earth globe, has the advantage of realism, over a flat map (mercator projection). For example, you need a globe to determine that Pt. Arena, California, is the closest point to Hawaii, on the west coast of America. The astronauts used a globe.

FLAT MAPS FOR A ROUND WORLD

It is impossible to have a totally accurate map of a three dimensional globe on a flat surface, but many have tried. The best efforts are as follows:

Map around Homer's time

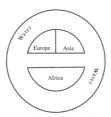

ROBINSON

Arthur H. Robinson in 1963, probably has produced the best map with the least distortion, although Greenland is too small.

VAN der GRINTEN

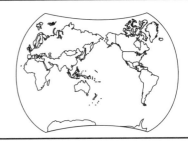

Alphone van der Grinten in 1904, helped to solve the extreme distortions of the Mercator projections at the high latitudes, but Greenland still is the size of South America.

MOLLYWEIDE

Karl B. Mollyweide in 1805, has the relative sizes of the continents correct, but higher latitudes are distorted.

MERCATOR

Gerhadus Mercator in 1569, gave us the most widely used flat map, but the high latitudes are greatly distorted. Mercator was the first to use the word "atlas", for a volume of maps.

Over 200 flat maps of earth have been produced and all are wrong.

When you flatten a globe into a map...the scale will be true along only one great circle, or two lines of latitude.

Robinson chose 38^{0}N and 38^{0}S, for his two latitudes.

New space maps for astronauts, use a great circle of earth, or a planet, over which they pass, for their mercator projection.

HOMOLOSINE PROJECTION

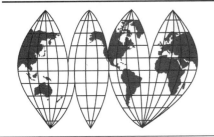

Although the oceans are distorted, by dividing, the map provides a minimum distortion of land masses. The map represents, "the peeling of an orange."

ANCIENT MAP

The ancients knew of the Tropics, arctic circles, equator, latitude, longitude and the ecliptic.

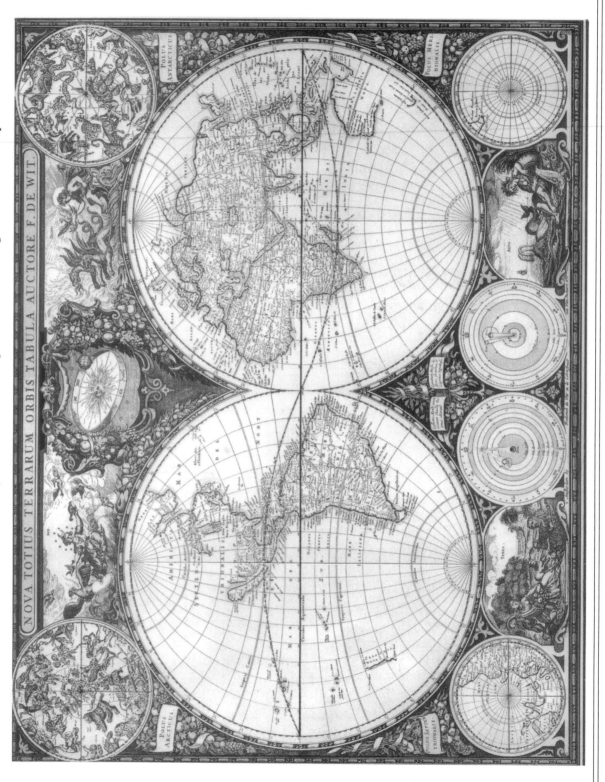

NOVA TOTIUS TERRARUM ORBIS TABULA AUCTORE F. DE WIT.

First-class West Indiaman *Thetis* (Captain Burton) getting under weigh off the Needles, Isle of Wight.

THE FOUR SEASONS

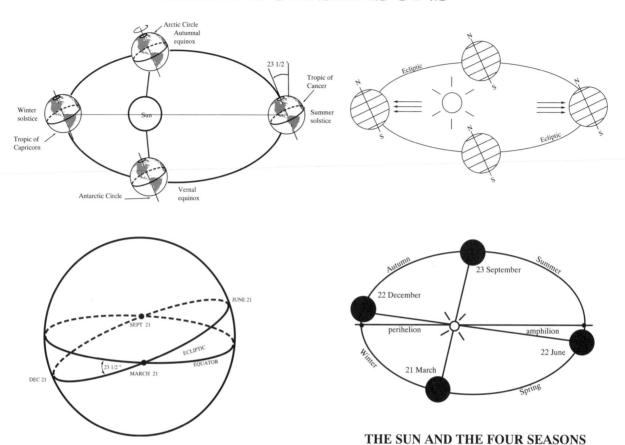

THE SUN AND THE FOUR SEASONS

The *perihelion* is the point in the orbit of the earth, planet or comet that is nearest to the sun.

The *aphelion* is the point in the orbit of the earth, planet or comet that is furthest from the sun.

Note that in the northern hemisphere, when it is winter, the sun is the closest to the earth. ***The earth is at perihelion. The tilt of earth causes seasons.***

The tilt of the Earth revolving around the sun causes the seasons, the solstices and the equinoxes. Note that the sun is closer to the northern hemisphere in winter.

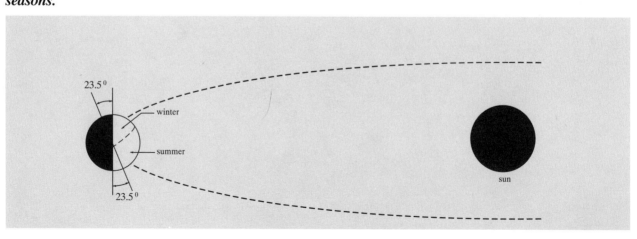

EARTH PRECESSION

(Pronounced: pre-ces-shun)

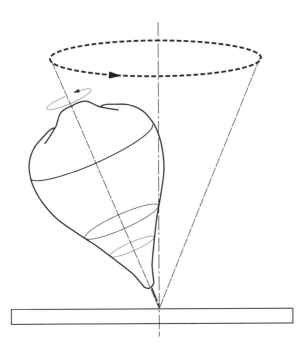

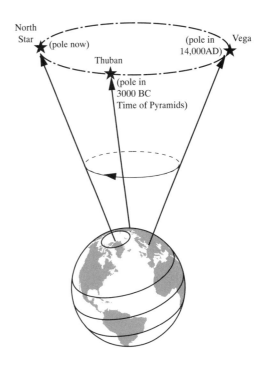

North Star (pole now) · Thuban · (pole in 3000 BC Time of Pyramids) · Vega (pole in 14,000AD)

Precession is the slow conical motion of the earth's axis of rotation, caused by the gravitational attraction of the sun, moon and planets.

Not to be confused with 'wobble', which is the small movement of the earth's axis (in inches) as it rotates.

EARTH MOVES

Look at your clock or watch and concentrate on the second hand.

As the second hand ticks and rotates, did you feel anything?
Did you sense any movement?
Every time the second hand moves, the earth travels 18 miles around the sun.
In a minute the earth travels the length of the State of California.

THE FIVE TERRESTRIAL ZONES

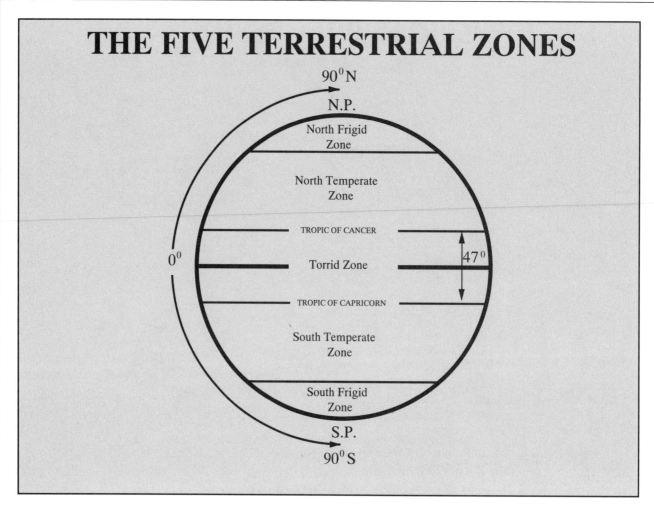

90^0N

N.P.

North Frigid Zone

North Temperate Zone

0^0

TROPIC OF CANCER

Torrid Zone 47^0

TROPIC OF CAPRICORN

South Temperate Zone

South Frigid Zone

S.P.

90^0S

STARS AS A GUIDE

Sailing with the stars as a guide, using the rigging.

Painting by Winslow Homer

Using the sextant and marking the time.

THE NORTHERN HEAVENS

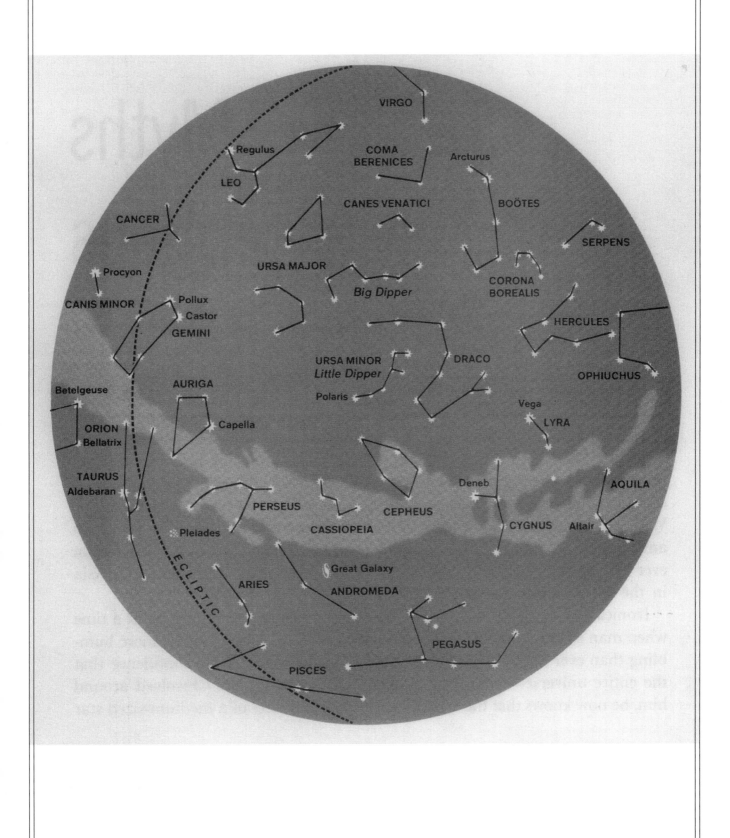

THE SOUTHERN HEAVENS

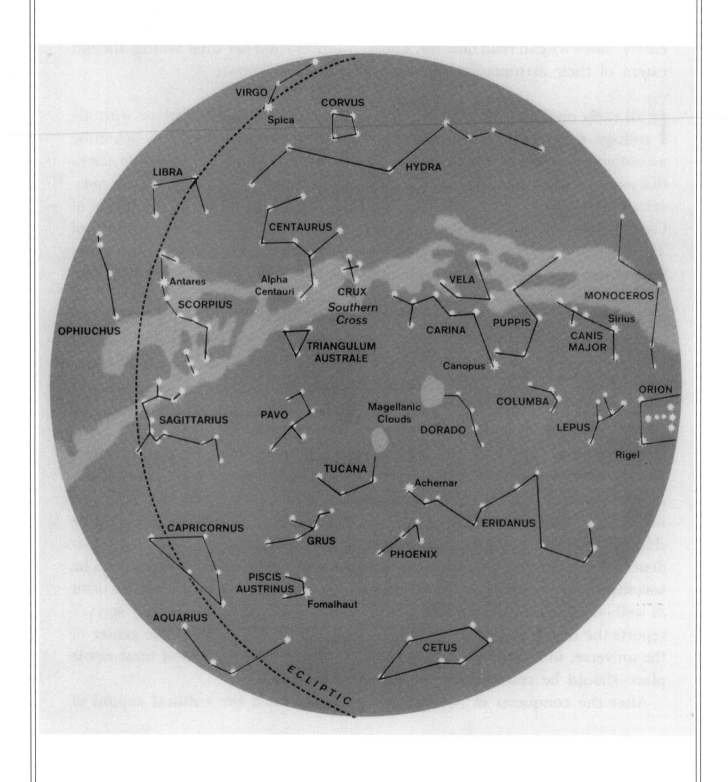

CELESTIAL SPHERE
WITH ECLIPTIC AND EQUATOR

The earth's orbit revolves around the sun.
Note that the ecliptic, the orbit of the earth,
is at an angle of 23 1/2° with the equator.
The ecliptic extended into space, represents
the earth's orbit...it revolves through all of
the signs of the zodiac.

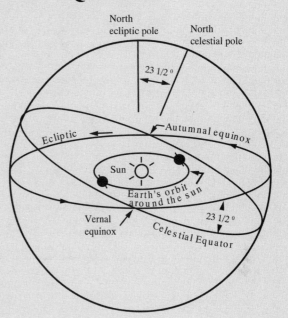

ECLIPTIC OR CELESTIAL ECLIPTIC
{ 23 1/2 °
EQUATOR OR CELESTIAL EQUATOR

THE "LINE OF THE NODES"

The moon's orbit around the earth, is inclined at
an angle of 5°, in relation the earth's orbit
(ecliptic).

The two points where the moon's orbit crosses
the ecliptic, are called the ascending and descending nodes. Note that the incline is 5° with
the ecliptic and not the equator. This makes the
moon's plane at a 28 1/2° angle, with
the equator.

The line connecting the 2 nodes, is called the
"line of the nodes."

It can be noted on the analemma, that the 2
crossings occur at 9°N latitude. It is believed
that this is a new definition.

You can observe on the analemma, that at 9°N,
the sun is ascending and then later in the year it
is descending. This is the only location on the

analemma, where the sun is at the same point,
during the year. It is the only point on the
analemma, where the sun crosses.

The plane of the moon intersects the ecliptic at
9°N, this is north of the equator.

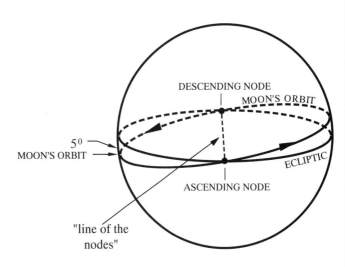

FIRST POINT OF ARIES

The equivalent of the brass plate at Greenwich.

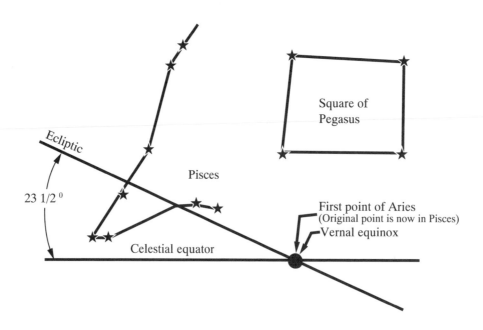

Square of Pegasus

Ecliptic

Pisces

23 1/2 °

First point of Aries
(Original point is now in Pisces)

Vernal equinox

Celestial equator

THE ECLIPTIC PLANE

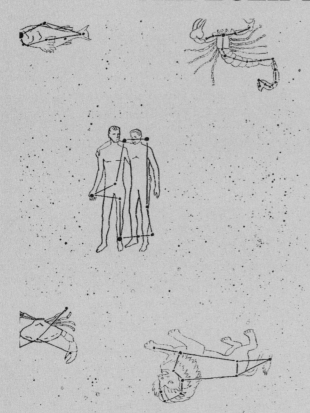

The ecliptic plane, extended to infinity, represents the revolution of the earth around the sun. It also represents the center line of our zodiac. The ancients knew the path of the earth through the constellations and gave the constellations fanciful names and designs.

MERIDIANS, ECLIPTICS & ANALEMMAS

The virtrual sun's latitude, plotted as points for 365 days, at 12 noon. for all meridians. forms an ecliptic.

Copernicus, followed by Galileo. showed us that it is a virtual sun, because the sun does not revolve around the earth, but in reality the earth rotates on its axis and revolves around the sun,

The path of the sun on earth and the ecliptic (orbit of earth)are identical. The ecliptic forms a

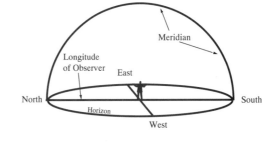

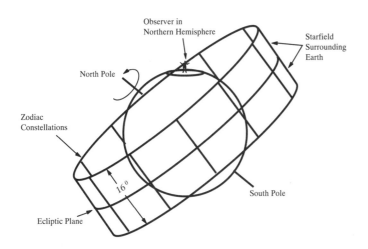

You can think of the path of the sun on earth and the analemma as equals. We believe this is a new way of looking at analemmas and ecliptics ... for any celestial body,

It is one key to the universe and celestial navigation.

circle around the earth and **all** meridians.

As this circle is expanded into the sky it forms the celestial ecliptic. It is also the ecliptic of the zodiac. It is very close to a circle, but is actually an ellipse.

The exact same logic as above, but plotted for **one** meridian, is an analemma.

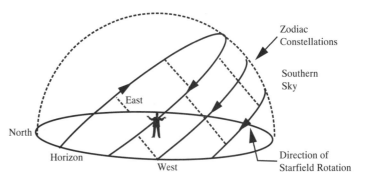

Frigate   Seventy-four   Pilot boat
At Spithead

A FEW POINTS OF INTEREST ABOUT THE SUN

1. With 360° in a circle and 24 hours in a day. The sun travels 15° in one hour, 360 divided by 24.

2. At the equator, or on any great circle, 15° is equal to 900nm (nautical miles). As mentioned 15°, measured with fist and thumb, or 1/2 of a hand spread of 30°, is the distance the sun will travel in one hour.

3. Knowing the above, it is easy to decide where to sit, when outside, to be in the sun, or shade.

4. If the sun is 15° from the western horizon, them the sun will set in one hour. Try it.

5. When the sun, or any star, is at zenith, being directly overhead, or directly over a position. It will be 90° from the horizon, The vertical and the horizontal form a 90° angle, or three hand-spans, equal to 90°.

6. The sun is an average of about 93 million miles from earth.
Fortunately, distances from earth to the stars, are not involved in celestial navigation.

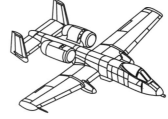

7. If a plane started on the equator, at 12 noon, and traveled at 900nm (nautical miles) per hour, and remained at the surface of the equator...it would always be 12 noon as it circumnavigated the earth. The plane would be keeping in exact movement with the arc of the sun, 15° per hour.

THE TERRITORY CURVES

The curvature of the earth is indicated in the illustration. At one mile the curve drops below the level by 8.4 inches, at three miles it drops 6 feet, and at fifty miles it represents a curvature with a drop of 1,668 feet. With your eye at sea level, you can not see a 6 foot man at 3 miles. This is the 3 to 6 rule.

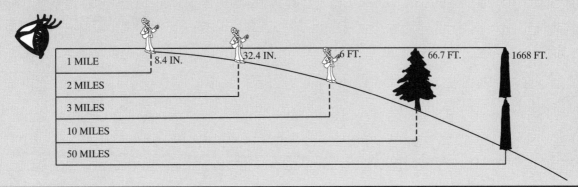

RHYTHM OF THE SPHERES

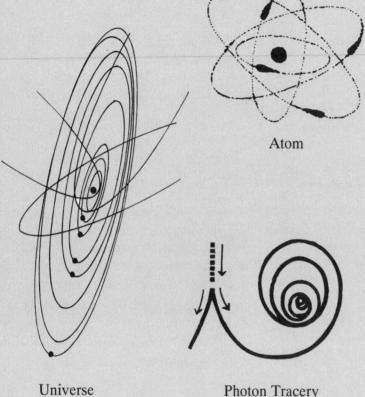

Universe

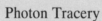

Atom

Cell

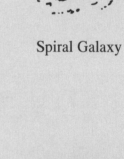

Photon Tracery

Spiral Galaxy

Nautilus Shell

Snow Flake

Double Spiral

TO MEASURE THE EARTH WITH A SHADOW

NOTHING CAN REMAIN IMMENSE IF IT CAN BE MEASURED.

Eratosthenes, a Greek poet, was the head librarian in Alexandria. He noticed that at high noon during the summer solstice (the sun's Zenith), the sun shone directly into a well at Syene (now Aswan).

He reasoned that because of the curvature of the earth, the sun would not be shining directly down in Alexandria located five hundred miles to the north (a twenty day's camel ride).

He waited a year until the next solstice, and then measured the shadow in Alexandria cast by a pole absolutely plumb (at high noon), The shadow cast at an angle of 7.2°, From simple mathematics he knew that a line crossing two parallel lines made corresponding angles, thus the angle from Syene to the center of the earth and back to Alexandria was also 7.2° He knew that 7.2° represented 500 miles, and that the full circumference of the earth represented 360°, or 50 times more than 7.2°, He estimated the circumference of the earth at 50 times the 500 mile distance, or 25,000 miles.

Modern science, using sophisticated electronic and photographic equipment, tells us that the equatorial (latitudinal) circumference is 24,904 statute miles, and the meridional (longitudinal) circumference is 24,860 statute miles.

Eratosthenes didn't do too badly considering his only tools were his brain and a stick. Also, he did it over 2,000 years ago.

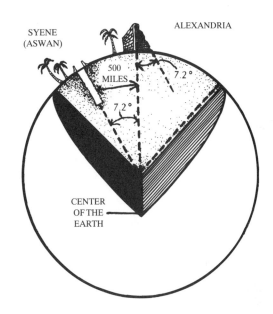

SYENE (ASWAN)   ALEXANDRIA

500 MILES   7.2°

7.2°

CENTER OF THE EARTH

Ancient man observed that it took about 30 days for the moon to travel around the earth (a 'moonth'). They also noticed it took 12 months for the sun to reach its zenith (solstice), as well as certain stars. The 12 months, or 360 days became the basis for our year, and also the reason we use 360 degrees in a full circle.

In future millenniums if we can eventually navigate
to the edge of the finite universe..it will present three questions:
1. What is beyond?
2. What do you call it?
3. How far does it extend?

MOON NAVIGATION

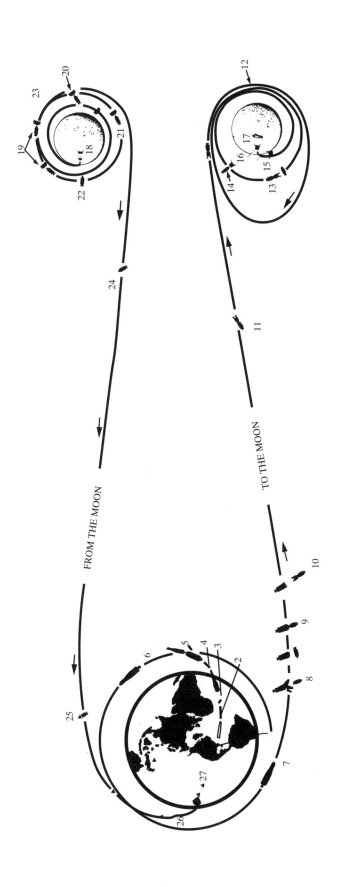

FROM THE MOON

TO THE MOON

1. Liftoff
2. S-IC powered flight
3. S-IC/S-II separation
4. Launch escape tower injection
5. S-II/S-IVB separation
6. Earth Parking orbits
7. Translunar injection
8. CSM separation from LM adapter
9. CSM docking with LM/S-IVB

10. CSM/LM separation from S-IVB
11. Midcourse correction
12. Lunar orbit injection
13. Pilot transfer to LM
14. CSM/LM separation
15. LM descent
16. Touchdown
17. Explore surface, set up experiments
18. Liftoff

19. Rendezvous and docking
20. Transfer crew and equipment from LM to CSM
21. CSM/LM separation and LM jettison
22. Transearth injection preparation
23. Transearth injection
24. Midcourse correction
25. CM/SM separation
26. Communication blockout period
27. Splash down

THERE HAS BEEN
AN ALARMING INCREASE
IN THE NUMBER OF THINGS
I KNOW NOTHING ABOUT.

MEASURING DISTANCE TO A STAR

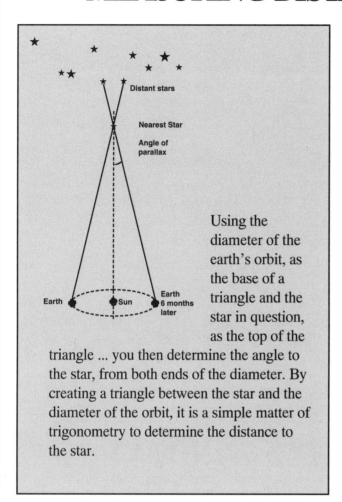

Using the diameter of the earth's orbit, as the base of a triangle and the star in question, as the top of the triangle ... you then determine the angle to the star, from both ends of the diameter. By creating a triangle between the star and the diameter of the orbit, it is a simple matter of trigonometry to determine the distance to the star.

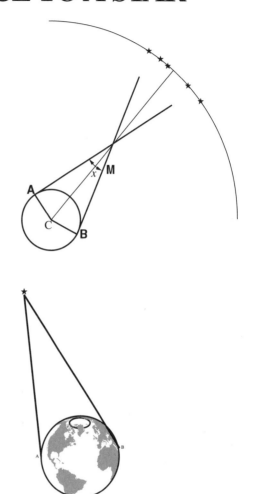

On a celestial sphere, Greenwich Hour Angle (GHA) is the same as longitude. It is measured westward from the Greenwich meridian from 0^0 to 360^0.

Also on a celestial sphere, right ascension is the same as longitude. It is measured from the vernal equinox, on the equator, or in the sky it becomes the First Point of Aries. It is always measured eastwards from zero to 24 hours.

The analemmas portrayed in this book are correct. The vernal equinox must be to the east of the analemma's meridian.

If the vernal equinox (VE) is to the west of the meridian, it is wrong and they have been shown on globes incorrectly for over 140 years.

The reason for this error is quite simple. Cartographers took the analemma. as it would appear in the sky and placed it on globes. This reversed the analemma and was wrong.

With coastal navigation, you use horizontal reference points. When you draw two lines from two coastal landmarks, where these two ship to shore lines intersect, is your geographic position, or fix.

Celestial navigation is similar in principle, but you are using vertical lines to the stars. With a sextant you form a circle of equal altitude. A second circle of equal altitude will cross the first circle. A third circle will indicate a point where all three circles cross ... this is your fix.

WINDS

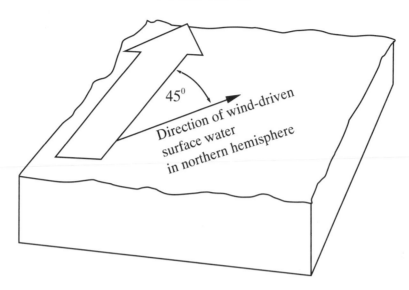

Wind drives the surface current at a 45⁰ angle to the direction of the wind. In the northern hemisphere it is 45⁰ to the right and in the southern hemisphere it is 45⁰ to the left.

THE EARTH'S SIX BAND WIND SYSTEM

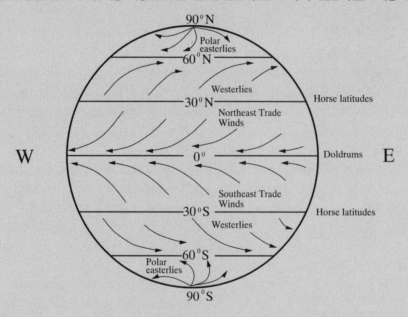

The rotation of the earth from west to east, creates the various wind systems.
Note that, at the equator, the wind goes clockwise in the northern hemisphere and
counter-clockwise in the southern.

DRIFTING LAND MASS

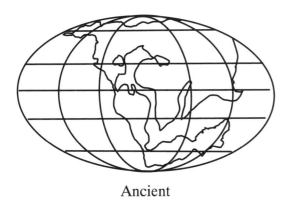

Ancient

Present

When the earth was first formed about 5 billion years ago, it was all one land mass, called **Pangea**. The continents gradually drifted apart to form our present land areas.

Carl Sagan, had a license plate that read, with a wry sense of humor, "SAVE PANGEA."

SEXTANT ANGLE

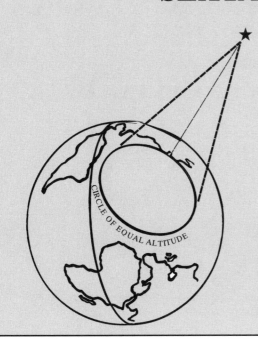

CIRCLE OF EQUAL ALTITUDE

Note the transparent cone coming out of the earth representing a "circle of equal altitude."

The illustration shows the sextant cone using the north star. The angle is the latitude.

By using two other stars and ...... two more sextant (angle) cones you have the fundamental principle of sextant celestial navigation, where the three circles intersect is the GP.

When using a sextant, you are measuring an angle (altitude) to a star, at a specific time. This angle represent a circle on earth, a "circle of position " . A cone, from the star as a point and to a circle on earth. You are somewhere on this circle.

If you then take another "fix" (to plot your position), you will have two circles and two points of intersection.
A third fix and a third circle, will give you a confirming point of intersection and your geographic position (GP).

A "line of position" (LOP) , can be circles of:
  equal altitude
  meridians of longitude
  parallels of latitude
  a compass bearing
  a river; a road
  a coastline or railroad tracks.

You need at least 2 LOP's to have a point of intersection. Assumed position is AP. It should be within 100nm.
Dead reckoning is DR. Dead reckoning position is DRP.

THREE CIRCLES OF EQUAL ALTITUDE
DETERMINE A FIX

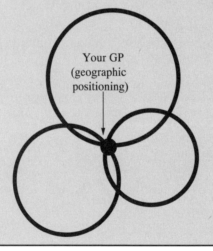

Your GP
(geographic
positioning)

CIRCLE OF EQUAL ALTITUDE

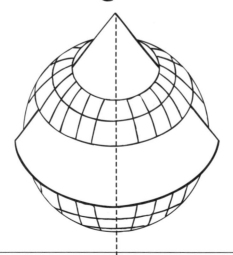

THE SIGNS OF THE ZODIAC

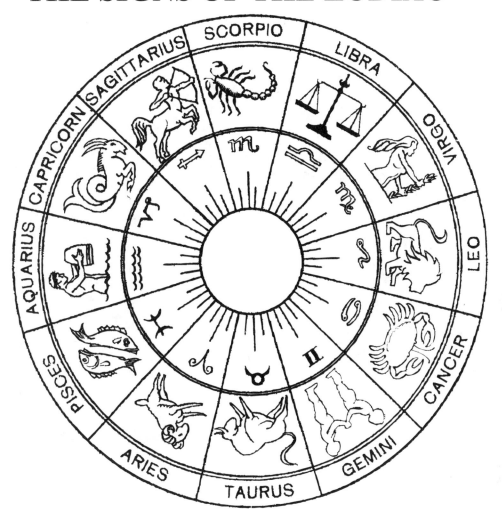

Aries	March 21 -	April 19	The Ram
Taurus	April 20 -	May 20	The Bull
Gemini	May 21 -	June 21	The Twins
Cancer	June 22 -	July 22	The Crab
Leo	July 23 -	Aug. 22	The Lion
Virgo	Aug. 23 -	Sept. 22	The Virgin
Libra	Sept. 23 -	Oct. 23	The Balance
Scorpio	Oct. 21 -	Nov. 21	The Scorpion
Sagittarius	Nov. 22 -	Dec. 21	The Archer
Capricorn	Dec. 22 -	Jan. 19	The Goat
Aquarius	Jan. 20 -	Feb. 18	The Water Bearer
Pisces	Feb. 19 -	March 20	The Fishes

The dates after the signs of the zodiac, represent the days
the sun is observed in the zodiacal belt, about 16° wide.

CONSTELLATIONS
AS THE ANCIENTS VIEWED THEM

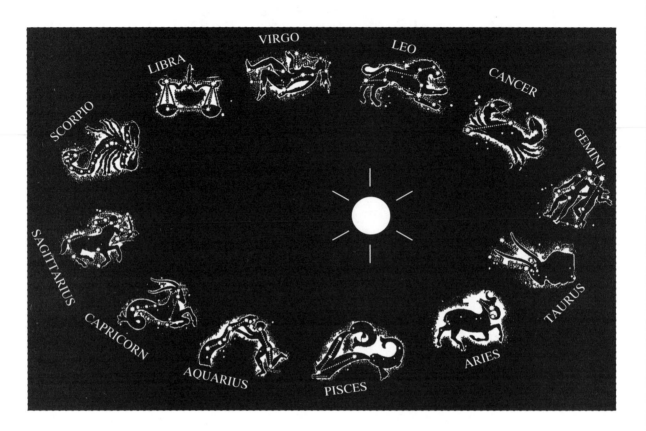

In the year 4000, it is not too extreme, to consider celestial navigators, born in space-craft, navigating in light years...never to return to earth.

On their way to the moon, the astronauts took sextant readings of the signs of the zodiac, to verify their computer positions.

"There is a big difference between a sailor and a navigator." Ernest Ruger

The "meridian of the analemma", can be any one, of an infinite number of meridians. By definition, you must select only one meridian, for one year. The latitude for the analemma is zero, or the equator.

However, the analemma could be observed at any latitude, the shape changing, as you move away from the equator.

If you understand the concepts of the analemma, it is believed you are in the top one per cent of the population.

Using a classified GPS, DOD code, you can plot your GP within inches.

The signs of the zodiac, are the names the ancients gave to the constellations (groups of stars). They observed the path of the sun through the heavens, or as we know today, the path of the earth as it revolves around the sun.

This path of the earth, or ecliptic, becomes the celestial ecliptic and is the median orbit, passing through all 12 constellations.

SUN'S POSITION IN THE ZODIAC

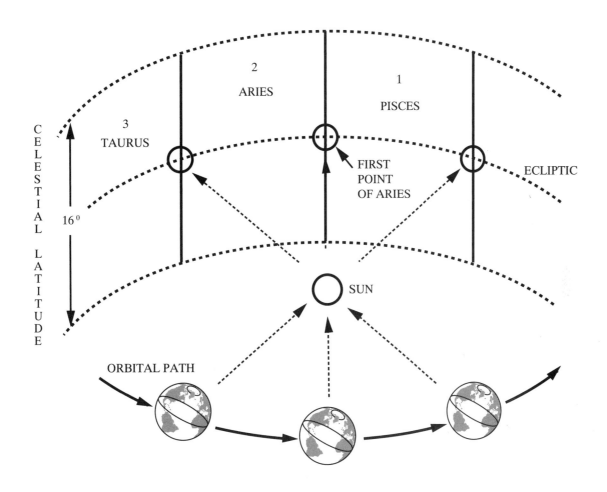

Note the ecliptic going through the center of the zodiac.

TILT DETERMINES SEASONS

Sun's Rays

TIME TO ORBIT THE SUN

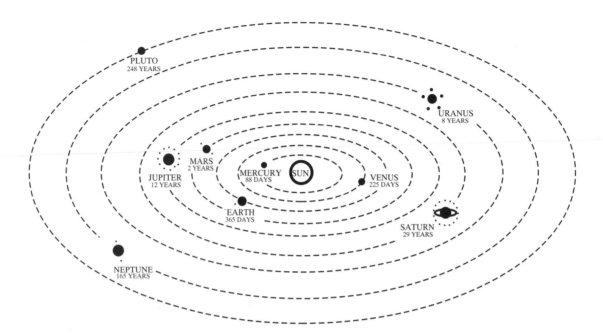

Solar system showing planets and their satellites, with the orbit of each
and the time, in round numbers, required by each planet
to revolve completely around the sun.

ROTATION

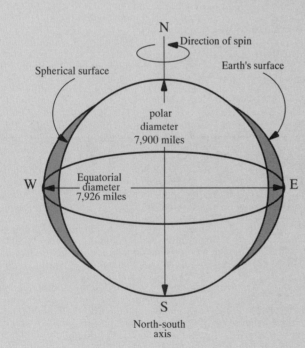

The earth rotating on its axis causes the center of the earth to bulge out.

The earth rotates in an easterly direction faster than the water can move. The interface between the surface of the earth and water, creates friction, because the water is lagging behind. This is called the *Coriolis* effect.

The friction between the water and the rotation of the earth, slows the earth about 1 1/2 milliseconds for each 100 years. The earth's rotation causes the equatorial diameter to be larger then the polar.

CELESTIAL EQUATOR AND ECLIPTIC

It is rather amazing, that two intersecting circles of equal size, can explain so many things. They can cover a lot of ground and sky.

Imagine a celestial equator, intersecting with an equal size ecliptic. What do we have?

1. We have the "ecliptic", the path of the earth around the sun.

2. The ecliptic is at a 23 1/2° angle, in relation to the "celestial equator."

3. The ecliptic intersects the celestial equator, at two points.

4 The two points of intersection are called the "equinoxes."

5. One intersection is called the "vernal (green) equinox", or "spring equinox."

6. The vernal equinox is called the "ascending node."

7. It is called the ascending node, because it is going from the "southern hemisphere" to the "northern hemisphere."

8. When you look at the vernal equinox on the "analemma", it is obvious that it is ascending.

9. The other intersection is called the "autumnal equinox."

10. The autumnal equinox is called the "descending node."

11. The autumnal equinox is going from the northern hemisphere to the southern hemisphere.

12. The ecliptic is in the median orbit of the "zodiac."

13. The zodiac is therefore, also intersecting the celestial equator at two points.

14. The point of intersection of the vernal equinox, in ancient times, on the celestial equator, was the "first point of Aries."

15. The first point of Aries, is today in the "constellation of Pisces."

16. The first point of Aries, is the "zero point", for the "measurement of stars."

17. The "signs of the zodiac" , on the ecliptic, are "precessing" (moving) to the west on the celestial equator.

About one million miles from earth, the gravitational pull of the sun, is equal to earth's.

Without our moon, the earth's tilt would wobble from 0^0 to 90^0. It is fair to say that the moon controls our seasons. It also stabilizes our navigation systems.

The diameter of the universe is estimated at 8,000,000,000 light years. A light year being 186,000 miles per second for one year.

FINAL EXAM

The shape of the analemma is determined by the tilt, of the axis of rotation, of the planet and the "equation of time", for that planet.

Final exam!

What is the shape of the analemma, on an unknown planet, revolving around an unknown sun?

You need more information to determine the shape. Add two conditions:

1. It rotates on its axis and has no tilt.

2. It revolves around its sun in a perfect circle.

See if you can guess the answer.

Stop here.

Answer:

Both the Tropic of Cancer and the Tropic of Capricorn are on the equator at zero degrees.

There is no width, because there is no time difference. There is no height, because there is no tilt.

Since the revolving around the sun is without distortion. There is no equation of time. The answer is that the analemma is a point.

For more testing try your knowledge: CONTENTS on pages 6 and 7.

For each mile to the moon,
it is about one million miles to Mars.
For each day to the moon,
it will take about one year to go to Mars.

APPENDIX

The following non sequiturs, are bon mots of information that relate to navigation. We hope they are of interest:

The longer the path of a planet, or celestial body, around any sun, the longer the time between the solstices.

On earth the Tropics and solstices, are determined by the tilt. If there is no tilt, the solstices of a celestial body are on the equator.

Different color lasers placed at the Prime meridian, solstices, equinoxes and north and south poles, of the planets...could be used as light houses for the universe and space travel. This will probably be obvious to the unborn.

If any celestial body rotates, around a sun, in a perfect circle there is no equation of time. A 12 noon fix, on any meridian, will be at the exact same time, as clock time.

An ecliptic, can be considered as an analemma for all the meridians, of any planet. We believe this is a new definition.

Or the reverse, the analemma can be considered as the ecliptic for one meridian.

If you know both the angle of tilt and the equation of time, for any celestial body, known or unknown, you can determine the analemma. Knowing the analemma, you know the ecliptic. As such, the analemma can be a key to navigation and one of the keys to understanding the universe.

The vernal equinox, represents many things at once: the point on the equator when the sun goes from the southern hemisphere to the northern hemisphere; represents a time of equal day and night; projected into space it becomes the "first point of Aries", the zero point for measuring star locations; and to the ancients, it represented the time to plant. Vernal meaning green.

To the west of the vernal equinox is the autumnal equinox, the point on the equator that represents the sun going from the northern hemisphere to the southern hemisphere.

When Einstein was asked what is beyond a finite universe, he responded, "Gott (german) only knows."

It took 1,400 years for astronomers to change from a geocentric (earth centered) universe to a heliocentric (sun centered) universe.

On the analemma, if you note the 4 quadrants formed by the equator (x-axis) and the meridian (y-axis), you have the 4 seasons of 3 months each.

The NW quadrant is spring; NE is summer; SW is fall SE quadrant is winter. These are the seasons of the northern hemisphere. You may want to check your travel plans with the position of the sun.

Other than at the solstices, the sun moves north or south, on the analemma and the ecliptic (orbit of earth), about 1/4 degree per day.

Knowing that the sun moves 15° per hour, you can estimate the zenith position...the highest point of the arc of the sun, that represents a true north/south line.

Dead reckoning is the way navigators pronounced "ded." Ded is the abbreviation for deduced reckoning.

TAD, acronym for: Time, Angle & Distance.

The plane of the equator is a great circle, at 0° latitude, that extends to infinity and becomes the celestial equator.

A planet rotating, in a perfect circle, around a sun, but with a 40° tilt...would have an analemma that is a vertical line, on its meridian, that extends from latitude 40°N to latitude 40°S.

A planet rotating, in an elliptical orbit, around a sun, but with zero degree tilt...would have an analemma the is a horizontal line, on its equator.

To visualize the analemma as it is on earth, you can imagine a transparent globe and you are in the center of the globe looking up. In this way you can see the analemma as it is and not reversed...as it has been incorrectly done for 140 years.

There are more stars in the heavens, than there are grains of sand on all the beaches, in all the world.

Found in the log of a Viking ship, "...pods (group) of whales that lie a day south of Greenland." They then knew they were near Greenland.

The sun (virtual) appears to move from east to west, at 15° per hour. Actually, it is the earth rotating on its axis from west to east, at 15° per hour. The earth is also revolving around the sun.

To remember a country on the equator, remember Ecuador, it means equator.

Time-lapse photographs have been taken of Polaris (north star), They clearly show a circle of stars, rotating around a hub, with Polaris as the center.

Perhaps the single most compelling reason for exploring space, is man's curiosity.

The idea of a flight to the moon, was written by Lucian of Samosata, a Greek, in A.D.160.

All through history there have been instruments to measure angles of the stars for navigation, including: a stick with notches, cross-staff, four-staff, astrolabe, sextant and now Global Positioning System (GPS).

Global Positioning System (GPS), is an electronic/radio system, using satellites for transmitting latitude and longitude to earth receivers. The satellite "knows" where it is in the sky - and electronically calculates the angle and distance to a transreceiver on earth. The direction, angle, time and distance can then be translated into a global position (GP).

All great civilizations understood the ecliptic, solstices and equinoxes. Perhaps it could be one definition of civilization.

The tilt of a planet, or celestial body, determines the solstices. The closer the orbit around the sun is to a circle, the less the equation of time.

Why is the analemma always seen on a globe, to the west of South America? The answer is simple... this is the best position, without covering landmarks.

The analemma can exist on any meridian. The analemma could also be placed on any world map, or on any map that includes the Tropic of Cancer and the Tropic of Capricorn.

A modern clock does not agree with sun time. Local Apparent Noon (LAN), or 12 noon by your clock, is only in agreement with the sun, at zenith, which is 12 noon for sun time...4 times per year (note the analemma). To keep your clock, at the same time as the sun, you would have to adjust it daily.

Clocks that had a special cam and gear, to keep sun time, are now collector's items. Clocks use mean time and not sun time.

The difference in mean time and sun time is called the "*equation of time*."

The world is always divided into two different days, at the International Date Line.

In the book "ADRIFT, SEVENTY-SIX DAYS LOST AT SEA" the author, Steven Callahan, was lost in the Atlantic ocean, on a small raft. He used three pencils to form a 20° angle and used this to check his latitude with the north star. He arrived safely on a Caribbean island.

Over a thousand years ago Viking navigators took large black ravens on their voyage. They are non-migratory, land-based birds.
When released at sea they fly directly to land. If they do not see land they will circle and return to the mast.

It is estimated that when they soar to 5,000 feet, they can observe land over 100 miles away. If they spot a mountain, they will fly to it from over 200 miles away. The Viking navigator followed the ravens and arrived at landfall.

THE THRILLING NATIONAL BESTSELLER
Adrift
Seventy-six days lost at sea
"A tale of courage and determination in the face of almost insurmountable hardship."
THE NEW YORK TIMES BOOK REVIEW
Steven Callahan

Robert Dietz, of the National Ocean and Atmospheric Administration, has said of the Global Positioning System (GPS), the radio signalling satellites that orbit the earth, "This is the most striking innovation in navigation since the compass for orientation, the sextant for latitude and John Harrison's chronometer for longitude."

For over 140 years, the analemmas on globes, up to this day, are wrong. The author noted they are backwards. The analemma on these pages are correct. Bernie Oliver, an expert on analemmas, gave the author credit for this observation. Mr. Oliver, developed the hand-held calculator at Hewlett-Packard and helped to establish the Search for Extral Terrestial Intelligence (SETI).

The global positioning system (GPS), is the most accurate navigational system that exists. If you have one in your ship, car or hand, it is the system of choice.

If you have no navigational system, it is worth understanding dead reckoning. It should also be of interest to understand the concepts behind the systems.

Although dead reckoning, in your head, can not compare with the accuracy of GPS, the ability to point in the right direction, with the accuracy of your finger, has merit when you have no idea where you are.

John "Old number 4" Harrison invented the modern chronometer for determining longitude. His life is an incredible story.

Astronaut Neil Armstrong, in London for a formal dinner with the queen and prime minister, gave a toast, " To John Harrison, the man who helped us land on the moon."

It would not be unwise, to note the analemma and make travel plans, knowing the latitude of the sun.

John Harrison

Cosmology explains how stars arrived... astronomy explains where the stars are and of what they are made... celestial navigation explains your location on earth. To understand all three, is the measure of your knowledge of the universe.

The most difficult space to conquer is the space between our ears.

We can not discover new oceans, unless we have the courage to lose sight of the shore.

The great circle of the equator and the ecliptic, are parallel at the solstices and at the largest angle (23 1/2°) at the equinoxes.

We don't see
things as they are,
we see things
as we are.

The Wolf, brig of war (late of the Royal Navy), making signal and laying to for a pilot off Dover.

Placing a stick in the ground, you can observe that the path of the sun is a half-circle or arc. The top of the arc, marks solar noon and forms a line with the shadow of the gnomon (pointer on dial), this line represents true north and south.

This is one way of determining the zenith of the sun. The virtual sun moves from east to west, the shadow moves from west to east, when you are north of the sun.

You can also note the four cardinal points of the compass.

Albert Einstein stated, " The eternal mystery of the universe is its comprehensibility."

Apollo 12 astronauts set up reflectors on the moon, for laser ranging and accurate to within 14 feet.
It was determined that the moon is moving farther away from earth, so perhaps gravity is weakening.

British biologists at the University of Manchester, discovered iron deposits in the lining of over 300 human noses. Similar deposits have been found in birds, bees and dolphins. It is believed this could be an atavistic nose compass.

The first clocks sent to the United States, were accompanied by sun dials, used to set the clocks. At that time, they could not dial for the correct hour.

The analemmas of all the planets, could represent a key to space navigation

One nautical mile is equal to one minute of arc, on a great circle.

Since the earth is spherical and not flat, dealing with celestial navigation involves a "navigational triangle".

The 3 points of the triangle are:
(1) one of the poles, north or south,
(2) your assumed position (AP), and
(3) your geographic position (GP).

When you are trying to solve the navigational triangle, you are doing a "sight reduction." It is comforting to know you can solve a sight reduction" , by simply pushing a button on a GPS system.

The "noon shot" is when the navigator calculates the angle of the sun, as the sun is at its highest point, its zenith.

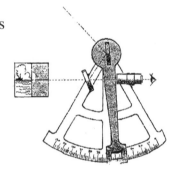

Using a sextant, navigators determine angles (altitudes) to: the sun, moon, planets and about 40 navigable stars.

The Polynesians and Micronesians used over 200 navigable stars.

The winter solstice is the shortest day of the year. The longest day of the year is the summer solstice.

Day and night are of equal length on the vernal equinox and the autumnal equinox.

I am not lost, I am investigating alternative destinations.

An emerald ring on a severed finger helped to speed the search for longitude. The unfortunate sailor drowned along with 2,000 others, off the southwest coast of England. Their ships crashed into the rocks, because the fleet navigator did not know the longitude.

The public were as outraged, as they were many years later, with the sinking of the Titanic.

If you are on a calm sea, or a flat desert, with your eye at sea level, you can not see a six foot man from a distance of 3 miles, even with binoculars. For the earth curves 6' every 3 miles.

Knowing this 6' to 3 mile relationship, you can calculate your distance from known landmarks.

The Vikings used "latitude sailing", as did the ancient sailors...up to the time of the discovery of the system for determining longitude.

The Vikings sailing west at night, kept the north star at right angles to the starboard beam. During the day they sailed parallel to the sun's path, on their port beam.

After Columbus, who also used latitude sailing, to discover the West Indies, there was a rule of thumb, for navigators: "Sail south until the butter melts, then sail due west until landfall."

Continuing to do the same things and expecting different results, is one definition of stupidity.

The Vikings, a thousand years ago, had navigation systems that are in use to this day.

---

*"And all I ask is a tall ship
and a star to steer by."*
John Masefield

If the wind will not serve take to the oars.
Latin proverb.

They used bearing or direction, by last sight of departure and first sight of arrival. They threw overboard a float or "log", to measure speed. The phrase "streaming the log", is still in use. At night they used a "husaotra", a vertical measuring stick, to determine the altitude (angle) of the north star and thus their latitude.

In daylight, the Vikings used our nearest star, the sun, as a compass. They were aware of the obvious east to west traverse of the sun, but also the north to south movement, that relates to the analemma, as were the Polynesians before them.

In 1751, George Washington, surveyed the island of Barbados. He was not only the father of our country, but also helped to father navigation.

Celestial navigators and astronomers, use a different language and frame of reference. The navigator wants to know where he is and looks to the stars for verification.

The navigator wants to know his latitude and longitude.

The astronomer knows exactly where he is. A 200 inch telescope, is not a portable instrument. It is hard to imagine any astronomer, lowering his telescope to the horizon and exulting "land ho. "

The navigator uses a fixed point on earth, to measure longitude. A brass strip at Greenwich, England. The astronomer uses a rather esoteric point in space, that is slowly but constantly moving.

It is called " the first point of Aries", the first sign of the zodiac and represents the exact time and point in space, where the sun crosses from the southern hemisphere to the northern hemisphere. It is the point on the earth's equator and the celestial equator, that is called the vernal equinox.

The navigator speaks of parallels of latitude, the astronomer of declination above and below the equator, which represents parallels of latitude extended to infinity.

The navigator speaks of longitude, translated from Greenwich Mean Time (GMT), the astronomer of Greenwich Hour Angle (GHA), which is GMT translated into an arc, that forms an angle.

The astronomer calculates "right ascension", which represents meridians of longitude, extended to infinity. Instead of using Greenwich as the zero point, astronomers measure from the first point of Aries and use time as the measurement, from zero hours to 24 hours. Measuring eastwards from the vernal equinox (the first point of Aries).

If your dead reckoning is accurate to within a minute of time, you can probably see your objective, or evidence of land. One minute of time, is equal to 15nm on a great circle. One minute of arc is equal to one nautical mile.

If you place a string on a globe, between any 2 points, it forms part of a great circle and is the shortest distance. Mark the distance of the string on the equator and each degree represents 60nm. A nautical mile is 1.15 statute miles.

One fathom is six feet. 1,000 fathoms is approximately one nautical mile. One league is 3 nautical miles.

Two elements necessary for any nautical fix are distance and direction.

When you draw an imaginary line, from the center of the earth to the celestial body you observe, where the line contacts the surface of the earth is the celestial body's geographic position, or "ground point" (GP).

Greenwich Time (GT) is the measurement of time measured on the prime meridian running through Greenwich, England. Also called: Greenwich Mean Time (GMT), Greenwich Civil Time, Universal Time (UT) and Zulu Time (military).

Greenwich time was made compulsory in England, by act of Parliament in 1880. The rest of the world has gone along with their logic.

"Only two things are infinite.
The universe and human stupidity
and I'm not sure about the former."
- Albert Einstein

The Victory, first-rate, 104 guns, In Portsmouth harbor , 1828; collier alongside.

Columbus in his later years speculated "the earth was not in the way that is usually written, but has the shape of a pear which is very round, except in the place where the stem is higher." He probably guessed, but he was right.

Satellite radio tracking indicates the north pole is about 60 feet above a symmetrical figure of earth and the south pole is 150 feet below. The earth's rotation causes the equatorial diameter to be larger then the polar diameter.

Satellite information also confirms the earth is "a pear shaped ellipsoid and includes large bumps and dips."

The strength of the earth's magnetic field averages about 50,000 gammas, one gamma is 10 to the minus 8th gauss, about the same strength as a toy horseshoe magnet.

Goethe said "You don't understand a thing, until you love it."

From ancient to modern times, the navigator with sextant and chronometer, was held in awe. The crew watched the "shooting of the sun" to determine position, on a boundless sea. It seemed more magic than science.

This is as the captain and navigator wished, for in denying the crew the knowledge of navigation and astronomy, it helped insure against mutiny.

Many countries hung sailors, who revealed the secrets of navigation.

About 500 AD, natives from the Marquessas, were first to Hawaii.

Time is nature's way of keeping everything from happening all at once and space is nature's way of keeping everything from happening in the same place.

In 1884 in Washington, D.C., Greenwich, England was selected as the prime meridian.

A clock aboard a space ship travelling at 87 per cent the speed of light, would run only half as fast as a clock on earth. Atomic clocks sent into space have recorded small differences in running time.

Archaeologists have found evidence at burial sites, stretching from Russia to Maine, of an ancient seafaring, circumpolar culture, over 7,500 years old. There is evidence in the burial artifacts, that they knew of triangulation.

This means that the concept of the modern sextant, was known in 5,500 BC.

The Polynesians and Micronesians, covered an area of the Pacific ocean, about 100° of latitude and 60° of longitude.

A sextant is worthless without an accurate measure of time.

In celestial navigation "angles" are referred to as "altitudes."

The horizon appears as a perfectly straight line and a star appears as a point.

Meridians of longitude are all great circles, are all 360^0, are not parallel, all are considered of equal size, and one degree of longitude is always the same distance.

Parallels of latitude are all 360^0 (except for the poles that are points), all are parallel, and only at the equator is the parallel a great circle. Only at the equator is a degree of latitude equal in distance, to a degree of longitude all of the other parallels are smaller circles and a degree of latitude is thus smaller in distance.

With all great circles, the equator, and with all meridians of longitude, one degree is equal to 60 nautical miles. It is an excellent point of reference.

Sixty nautical miles times 1.15, is equal to 69 statute miles.

Although the degree distances may vary, the same logic of latitude and longitude, applies to all planets, known or unknown.

One nautical mile per hour is equal to a speed of one knot.

The complete 360° cycle of the "precession of the equinoxes" takes about 25,000 years. This movement around the celestial ecliptic, this "cycle of precession" goes through the entire zodiac. There are 12 "signs of the zodiac", that represent 12 constellations.

It is caused by the precession of the earth's axis of rotation.
You will recall that the north star (Polaris), was not at the latitude 90°N, at the time of the building of the pyramids.

Look Pluton! they would have an analemma.

Yes Orion. but ours is different.

As the earth precesses, so do the equinoxes.

In the year 2600 BC, the pole star was in the Draco constellation. In the year AD 3500 the pole star will be Cephi.

The earth's ecliptic extended to infinity, is the celestial ecliptic. The celestial ecliptic goes through the center, of all the signs of the zodiac. The zodiac forms a band around the sun and earth, with a celestial latitude of 8° north and 8° south, for a total of 16°.

The first sign of the zodiac is Aries.
It is one of 12.

A sidereal (star) year also known as: astronomical year, lunar year, equinoctal year, solar year and tropical year is a division of time equal to: 365 days, 5 hours, 48 minutes and 46 seconds. It represents the time for earth to make one revolution around the sun.

The earth rotates on its axis and revolves around the sun. These are the correct terms to use.

The precession of the equinoxes is the slow movement of the vernal and autumnal equinoxes, along the celestial equator, the equinoxes in turn, pass through the signs of the zodiac, from east to west. At present we are in Pisces and then as the song says "the dawning of Aquarius", in about a thousand years.

The earth is revolving around the sun at a speed of 66,000 miles per hour. This is 30 times faster than a bullet.

Today a digital quartz wrist watch, is more accurate than the best chronometer of 60 years ago.

Compass measurements are over 60 times cruder than a sextant.

H.M.S. Prince, first-rate, 110 guns.

Maps were made with Paris, Canary Islands, Rhodes, Amsterdam, Munich, Brussels, Copenhagen, Rome, Warsaw, Madrid, Ferro (Germany), Christiania (Norway), Rio de Janero, Stockholm, Washington, D.C. and other cities, as the prime meridian, but England had the most ships, the best chronometers and the best navigators. The result was Greenwich, England, as the zero measuring point, for longitude.

If the zodiac was reduced to the size of earth, it would follow the ecliptic on a globe.

Over a thousand years ago the Vikings used the polarized direction of light as a compass.

Other than on the sextants used by SAS (Scandanavian Air Service) navigators, it is believed no other navigators used this technique, either before or since.

On the south east coast of Norway, is the town of Kragero. On a cliff jutting out into the fiord, is the source of the crystal, called cordierite. It is also known as iolite, water saphire and the Viking sun stone. Faceted and polished, it has been sold as ersatz saphire.

It is named after, Pier A. Cordier, 1777-1861, a French geologist. It is trichromic, having three colors, i.e., straw yellow, light blue and deep blue. It is the crystal, Eric the Red, used to help find the Americas.

When you scan the sky with the Viking sun stone, even in a pea soup fog, it turns deep blue when at right angles to the sun, even if the sun is 6° below the horizon. The author has verified that it works.

Einstein has told us that space curves, due to mass. The greater the mass the greater the curvature. Even light curves in space.

For navigation, we can use straight lines from celestial bodies to earth. The only curvature is the curvature of a planet.

"If truth were self-evident, eloquence would not be necessary." Cicero.

A simple test to see if your hand-span is equal to 30°, is to measure from the horizontal, to the vertical, It should be exactly 3 hand-spans, or 90°.

The ancients knowing that the sun going south, meant colder weather. They had celebrations, and sacrifices, at the time of the winter solstice (about December 22nd), to insure the sun would return north. It seemed to work.

If you know the tilt and the equation of time, for a planet, known or unknown, you can draw an analemma. If you know the number of days, for the planet to revolve around its sun, you will then know all of the dates on the analemma.

If there is life on other planets, there is an excellent chance they would understand the concepts of celestial navigation and the analemma, it could represent a common ground.

Without celestial bodies, there could be no navigation at sea.

If the axis of rotation of a planet, has no tilt, there are no seasons.

One seaman, knowing the latitude of the sun, from the analemma, steered a course to the west at 20°N latitude, and landed safely in the Hawaiian Islands.

Ferdinand Magellan (1480-1521), by sailing around the earth, physically proved it was round.

If you know the diameter of the planet, you then know the circumference, you can easily translate the 360° in a circle, to nautical miles per degree. You then have the distances, in nautical miles, for any great circle on that planet. This is a good start on space exploration.

The problem of longitude, became particularly critical after the first voyage of Columbus. In 1493, Pope Alexander VI, issued his Bull (papal document) of Demarcation, which was initiated to settle the squabbles, between Spain and Portugal, about who owned what land.

On a world map, His Holiness, simply drew a line from pole to pole, 100 leagues (about 300 statute miles) west of the Azores. Spain to have all land west of this line and Portugal to the east.

It was a Solomon like decision, but with one problem, nobody had any idea where the line was. 100 leagues west of the Azores, meant absolutely nothing, in terms of longitude. Today, we would know it as about 30°W longitude.

The search for a position fix, continued for several centuries.

Then on a foggy night, in October 1707, the English fleet was returning from successful battles against the French and Spanish, in the Mediterranean. They became lost in the fog and crashed on the rocks of the Scilly Islands, about 20 miles west of the southern tip of England. 2,000 seamen drowned. Rear Admiral Sir Clowdisley, Although very obese, swam to shore and as he lay exhausted on the beach, was murdered and robbed.

The loss of life and murder, so outraged the English, that Parliament, formed the Board of Longitude and offered a reward, to anyone who could find a method, to accurately determine longitude. The key to longitude was accurate time.

Red sky in morning, sailor take warning.
Red sky at night, sailor's delight.

After an incredible 50 year struggle, against the Board of Longitude and the complexities of making an accurate chronometer, John Harrison, the man who found longitude, developed with his fourth timepiece, "old number 4", an error of less than a quarter of a degree, or less than a tenth of a second per day, on a trip to Barbados and back. The author has a screen play on same.
An error of less than 15 seconds on a five month voyage. Both Captain Bligh and Captain Cook, used replicas of "old number 4."
With the help of King George, an amateur astronomer, he finally received his total reward, in 1773, when he was 80 years old.

Azimuth is a specific direction.

For thousands of years civilizations have been planting at the vernal (green) equinox and harvesting at the autumnal equinox.

Our Milky Way galaxy is like a cannibal. It devours other smaller galaxies and grows larger. Exactly how many others is not known.

Stephen Hawkings explains that our universe started with a "singularity"...a point of infinite density. It would seem in logic that the infinite density of a singularity, expands to infinite space.

Today's unthinkable is tomorrow's convention.

Black holes appear to form in the center of galaxies. A black hole pulls in all surrounding matter.

Galactic jets can spew out of a black hole...some scientists believe the jet is going faster than the speed of light.

Physicist Dr. Borge Noland, at University of Rochester, has speculated that if the universe was asymmetrical at creation, it raises the possibility of another universe being created with an opposite twist.

The United Kingdom (Edinburgh), steam vessel, 1000 tons burden, 200 horsepower .

Astronomical year is the division of time, between one vernal equinox and the next.

A leap year, is a year on the Gregorian calendar, that has 366 days, with February 29th as an additional day. It occurs in years, whose last two digits, are evenly divided by four.

For 361 days per year, the sun is not at zenith, at 12 noon, by our clocks, for any one meridian. The astronomical year, or sidereal year, never agrees with our Gregorian calendar.

The sun time and the star time are not "wrong", they are just different from our clock time.

Man makes clocks to run very accurately and to give us GMT. Man did not make the sun and stars, they do their own thing.

We adjust sun time, to clock time, with the "equation of time", in order to navigate. We adjust astronomical time to our calendars, with leap year. An astronomical year is different from our calendar, about one day longer, every four years. This is because it takes the earth, a little longer to revolve around the sun, than our calendar indicates.

When it is 12 noon at Greenwich, it is 12 midnight at the meridian, exactly opposite, which is longitude 180°. Longitude 180°, is also the International Date Line.

The International Date Line, is an imaginary line, at longitude 180°. All regions to the east are one day earlier, than regions to the west. Therefore, when it is Monday to the east of the line, it is Tuesday to the west of the line.

There are 24 hours in a day and 24 time zones on earth. As above, as Monday circles the earth, it disappears at the line, and Tuesday follows behind and then also disappears at the line. So we always have two different days on earth. As the earth rotates, Monday would start at the International Date Line, and also end at the line, Monday would have made one westward rotation, around the earth, in 24 hours. Following directly behind is Tuesday and then Wednesday, etc.

One day begins the other ends...at the International Date Line.
It should be noted that the International Date Line does not exactly follow the 180° meridian. The line has to go around countries and island groups, that do not want to be divided into two different days.

There are no gaps in either time, or days. After Monday has rotated 360° and returned to the International Date Line, Tuesday is right behind. The last second of Monday, at the line, is followed by the first second of Tuesday.

There are 400 billion stars in our galaxy, and there are 50 billion galaxies.

Absence of evidence, isn't evidence of absence.

If you reduced the sun to the size of a grain of sand; you could then circle the solar system with your arms. The nearest star would still be over four miles away.

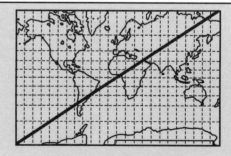

Rhumb line, is a line on a curve, such as earth, that cuts all meridians at the same angle. It is the path taken by a vessel or aircraft that maintains a constant compass direction.

During the vernal and autumnal equinoxes, the sun rises and sets on true east and true west., There are therefore only two days when this happens. The closer the date is to the equinoxes the closer to true east and west.

The sun always rises in an easterly direction and sets in a westerly direction, but only on the equinoxes is the sun on a true east to west course. By taking a right angle (90°) to this course at high noon...you will obtain true north and south.

The Danish physicist Niels Bohr, in 1913, calculated that the speed needed to prevent the electron of a hydrogen atom from colliding with the proton, was a revolution of seven billion times per second. This exact same logic applies to the revolution of the earth around the sun. If the earth stopped revolving it would end up a cinder inside the sun.

Can you name The Seven Seas of the modern era?
They are: Arctic, Antarctic, North Atlantic, South Atlantic, North Pacific, South Pacific and Indian.

One of the most recent theories, tells us that a body about half the diameter of earth, called Orpheus, collided with earth. This collision resulted in our moon.

When this primal moon first orbited the earth, it looked 15 times larger than it does today.

Some scientists believe the moon acted as a barrier, preventing celestial bodies from colliding with earth.

Gravity pulls together cosmic debris, that burns for billions of years.

In three billion years our sun will expand and turn the earth into cosmic dust...to form other stars and perhaps other life.

To measure the great distances of our universe in miles, astronomers use the astronomical unit (AU). An Au is the mean distance between the center of the earth and sun. It is about 93 million miles.

Note that the time zone on the meridian opposite the Greenwich meridian is centered on the International Date Line.

On a clear dark night, a person with good vision can see about 2,500 stars.

Einstein correctly predicted that the gravitational pull of the sun, would bend star light, 1.75 seconds of a degree. To visualize 1.75 seconds of angle, imagine a triangle with the base on a penny, and sides that extend 1 1/3 miles. Einstein's great thought was verified during an eclipse in 1919.

The sun is revolving clockwise in a nearly circular orbit, around the center of the galaxy, at about 200 miles per second. Making one revolution in 250 million years.

The azimuth of a star can be measured in degrees, along the horizon. The azimuth of a star directly to the north is 0°, 90° to the east, 180° to the south, and 270° to the west.

The sun contains 99% of the mass of our solar system, and is the most powerful gravitational force.

The light we see tonight, left Quasar OQ 172, 15 billion years ago, before our earth existed.

The nearest star is Alpha Centauri A. It is 4.3 light years away. A light year is the distance traveled by light (186,000 miles per second), in a vacuum, for one year. It is about 6 trillion miles.

Stars are born of other stars.

The ancients knew of solstices and named them after the sign of the zodiac that the sun was in at the time, i.e., The Tropic of Cancer and the Tropic of Capricorn.

To become a modern celestial navigator, you must know the terms of the trade. It is not that difficult to go from latitude and longitude, to a celestial sphere in the heavens.

Just visualize the parallels of latitude and the meridians of longitude, extending into the stars, to infinity. You are creating, in your mind, larger and larger spheres. When you have a sphere, large enough to show the stars of interest, you have a celestial sphere.

You can think of it as a large transparent globe, with the stars on the inside, of this sphere. For navigators, the distances to the stars is not required.

You simply use new names when dealing with a celestial sphere.

Declination of the stars, is exactly the same as latitude on earth.

Right ascension, in hours, minutes and seconds, is exactly equal to longitude.

You simply use 2 new terms, for what we already understand as latitude and longitude.

If you have followed the explanation of latitude, which is the number of degrees north or south of the equator...then you already understand the declination of the stars.

Declination, is simply the angle of the star in degrees. Declination of the star relates directly to latitude.

We have reviewed at length longitude. Longitude is the distance in degrees, east or west, of a brass strip at Greenwich, England.

There are no brass strips in the sky. So we need to create a zero point in the sky. We need a zero point to measure from. The vernal equinox was selected, by the ancients, as a starting point.

If you draw a line from the center of the earth, through the vernal equinox, on the equator, and extend the line to the celestial sphere, you have a zero starting point in the sky.
It is imaginary, but it works.

It is called "the first point of Aries". It exactly corresponds to the brass strip, at Greenwich. It is the zero celestial point, on a celestial equator for measuring longitude...except celestial navigators and astronomers, call it "right ascension."

Right ascension is measured in hours, minutes and seconds, moving to the west, from the zero point. I bet you could guess that one hour of right ascension, is 15°. You would be correct.

Capt. Bligh, of MUTINY ON THE BOUNTY fame, used a replica of John Harrison's "old number four" watch, to navigate thousands of miles back to England. in an open boat.

One advantage of a globe, is that with a piece of string, representing a known distance, you can determine any distance on the globe. It can readily indicate the shorter polar distances following a great circle.

The globe is the only way to show that Point Arena is the shortest distance from the North American continent to Hawaii. Astronauts used globes.

 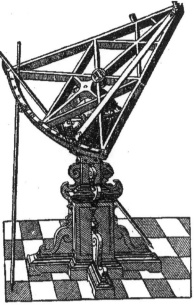

Tycho Brahe at Uraniborg observing with his great mural quadrant (left).

On the Right is a Sextant used to measure angles between celestial objects. (Rare Book Division, The New York Public Library, Astor, Lenox, and Tilden Foundations)

When it is 6pm GMT, (1800 hours), at Greenwich, it is 6am GMT, (0600 hours), at the International Date Line. It is also 12 noon GMT, (1200 hours), in Chicago and 10am GMT, (1000 hours), in California. These daylight hours, from 6pm to 6am, represent 180° of daylight.

These daylight hours do not represent a change in the day...the day changes only at the International Date Line.

The equator acts as the bow of a ship slicing through the water, half to starboard and half to port...except with our atmosphere, half the wind goes to the north and the other half to the south. This rotation explains why tap water in the northern hemisphere goes clockwise and in the southern hemisphere goes counter-clockwise.

When mariners were becalmed, they had to save water and threw their horses overboard...hence the name "horse latitudes."

Perigee is the point in the orbit of the moon, satelite, or heavenly body, when it is closest to earth.

Apogee is the point in the orbit of the moon, satelite, or heavenly body, when it is farthest from earth.

The sun is about 93,000,000 miles from earth; the moon is about 240,000 miles from earth.

The Navastar Global Positioning System started in the 1980's. The system allows astronauts to store their course in a computer. They can then verify location and speed, within a few feet.

Every point on a given latitude, is the same distance from the equator, the north pole and the south pole.

The east and the west 180th meridian are identical.

Space navigation is also called *astronavigation*.

At an altitude of 100 miles, satellites can orbit the earth. This can be considered, as one definition, of the beginning of space.

Sailors should carry large zip-lock bags. They are excellent for storage and can also be used under clothes, as emergency life preservers.

A plumb bob, or a weight on a string, was known by the ancients, to determine the vertical, or right angle. The Egyptians, also used grooves filled with water on the top of stones, to determine the level.

In the 17th century, Johannes Kepler, in Germany, accurately described the orbits of the planets. Astronauts use his calculations.

The plaque on the moon's date is incorrect. A.D. should precede the date

The celestial equator is also known as the *equinoctal* circle.

Intergalactic space, is the space between the galaxies, of the universe.

On a mercator (flat) map, the ecliptic looks like a squiggly line, going around the equator. More accurate is to consider the plane of the ecliptic, as a great circle, going through the center of the globe at a 23 1/2° angle to the equator.

The right ascension is exactly the same, as the measurement of time and degrees for longitude, on earth. The only difference is, that we are doing it on a celestial sphere

Right ascension, goes from the zero point, or zero hours to 24 hours, 24 times 15° per hour, equals 360°, a full circle.

For a quick summary...the X-axis, or latitude on earth, is declination on a celestial sphere.

The Y-axis, or longitude on earth, is right ascension, in hours or degrees, on a celestial sphere.

You now have the concepts of celestial navigation. Not that difficult, when you take a little time to think.

THE COLUMBUS ERROR

Columbus relied on the circumference calculations of Strabo (c. 63 B.C.) at 18,000 miles, he should have used Erastosthenes calculation of 25,000, who lived some 150 years before. As the world was 7,000 miles longer than Columbus thought...he believed he landed in the Indies, so chief Sitting Bull was known as an American Indian

When it is 12 noon, GMT, or Zulu time, in Greenwich, England, at the International Date Line, it is 12 time zones later (+12), to the east.

At the International Date Line, to the west, it is 12 time zones earlier. To our eye, the sun travels from east to west, so the sun has yet to arrive, to the west.

It is therefore, 12 midnight, or 2400 hours GMT, or Zulu time, at the International Date Line.

It is also, 1am, 0100 hours, at the time zone, just to the east of the International Date Line.

Greenwich, England has 12 time zones to the east and 12 time zones to the west, for a total of 24 time zones.

The sun (virtual) takes about 24 hours to circle the earth.

Whatever the time at Greenwich, it is the exact same number at the International Date Line, or 180°. Greenwich at 12 noon, is 12 midnight at 180°.

Rather than looking at the signs of the zodiac, future astronauts will be in them.

To avoid confusing timetables, when scheduling around midnight, it is preferable to use 23:59, or 00:01, so you can determine which day it is.

We have looked at determining your GP, by knowing the latitude of the sun, as shown on the analemma...then measuring your horizontal distance in degrees, north or south of the sun...measuring from an imaginary vertical line, to the east or west, from your location, to an imaginary vertical line from the center of the sun.

The reverse is also true. If you know the latitude, at your location and the number of degrees you are from the sun...you can determine the latitude of the sun, for that day.

By knowing the latitude of the sun, or its zenith point, and parallel of latitude, for that day. You can be quite certain that it will be hot at that latitude.

To see the world in a grain of sand
And a heaven in a wild flower
Hold infinity in the palm of your hand
And eternity in an hour.
William Blake (1757-1827)

Jean Foucault suspended a 200 foot pendulum weighing 60 pounds from the ceiling of the Pantheon in Paris. A stylus attached to the pendulum cut a path into a layer of sand placed on a table. It was found that the pendulum's oscillating plane seemed to rotate. Since the force of gravity was the only force acting on the pendulun downward direction, the rotation demonstrated that the earth itself is rotating.

The ecliptic (orbit of earth) intersects the celestial equator at two points. These are the two equinoxes. They are clearly shown on the analemma. The two intersections of the plane of the moon's orbit with the ecliptic, are indicated on the analemma at 9°N.

Dick Gordon, astronaut, has pointed out, that most photos of the moon are placed upside down.

Where the meridian of longitude, intersects with the parallel of latitude, is your geographic position (GP). That's navigation! That's it!

As with bacon and eggs, you must be committed to navigation.
The chicken is involved, the pig is committed.

On your globe, if March is to the west of the analemma meridian, it is wrong.

Thuban, was the north star, 5,000 years ago. Egyptian pyramids have shafts directed to the then north star.

We see stars backwards in time, as it was when the light left the stars...for many it was billions of miles away.

As Hubble said, "The stars as we see them tonight, have never existed."

Two points make a line; three points make a plane.

Einstein said, "God does not play dice with the universe." Stephen Hawking said, "God not only plays dice with the universe, but sometimes throws them where they can't be seen."

"What we have learned is like a handful of earth; what we have yet to learn, is like the whole earth." Avya

Light scattered and bent by the atmosphere, causes stars to twinkle.

The known universe is expanding every second, at a volume equal to our entire Milky Way galaxy.

Einstein made the following observation, "A theory is more impressive, the greater is the simplicity of its premises, the more different kinds of things it relates and the more extended its range of applicability."

The European Space Agency (ESA), uses a ring laser sextant, for spacecraft guidance.

Circa 450 BC, Oenopides is said to have discovered the ecliptic.

Claudius Ptolemy (circa AD 85-AD 165), a Greek astronomer, from Alexandria, Egypt, proposed the earth as the center of the universe, or geocentric theory. His ideas prevailed for 1400 years.

A Polynesian navigator would lie in the bilge of his canoe and interpret the slap of the waves against the hull. Knowing the angle and direction of the swells, he could locate an unknown island.

The Andromeda galaxy is 2 million light years away. If we had two identical twins on earth, both 20 years old and one was an astronaut and departed for the Andromeda Galaxy, at the speed of light...on his return to earth he would have aged 56 years and be 76 years old. The other 20 year old twin would have been dead for four million years.

Einstein tells us that traveling at the speed of light slows the aging process.

This is the BBC, it is 1600 hours Greenwich time (tone sounds). In California it is 8am, eight time zones to the west.

Based on recent progress reports...it may be a Japanese crew that first lands on Mars.

On November 1, 1883, the United States used GMT to determine local time.

On November 1, 1884, the International Meridian Conference, in Washington, DC, established the 24 time zones.

THE NOON MERIDIAN

When it is noon on a point on the meridian. It is noon on all of the meridian.

The polynesian navigator knew the path of the star Arcturus, passed directly over the Hawaiian Islands (just south of the Tropic of Cancer), and he knew the star Sirius, passed directly over Raiatea, in the Society Islands.

By sailing upwind of the island, under the path of the star, he could sail due west and make landfall.

Without any modern navigational instruments, he could find an island in the vastness of the Pacific, an ocean four times larger than the United States.

Other clues the navigator used were: water changing from blue to green, indicated a reef; driftwood indicated an island to windward; birds fly to land; estimate drift due to wind, by the angle of the canoe to its wake; the compass was the position of some 200 stars, rising and setting on the horizon; aligning with landmarks on his home island and noting any drift; and a large cloud in the distance, represented an island underneath.

The navigators could relate even the smallest clue to their position. A bird flying with a small fish in its beak, is nesting and flying to land, to feed its young. Follow the bird.

Captain Cook observed of the Hawaiians "They sailed with the sun serving them as a compass by day and the moon and stars by night."

They discovered islands in the Pacific, long before Columbus or the Vikings set sail.

Scintillate, scintillate, globule vivific.
Famed would I fathom thy nature specific.
Loftily poised in the ether capacious,
Strangely resembling a mineral ignescious

Half of all the knowledge of the universe has been published since 1985.

ASTRONAUTS

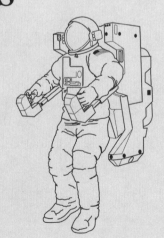

"Ride'Em Cowboy", Pete Conrad to Dick Gordon as Dick was attaching the tether to the Agena Rocket on Gemini II.

Richard F. Gordon, Jr., Command Module Pilot, on Apollo 12, the second landing mission to the moon. Completing the crew were Charles Colnrad and Alan Bean. Despite their Saturn V Rocket being struck by lightning 36 seconds after lift off, temporarily interrupting the electrical system, the mission was a success.
On Gemini II Dick went on 2 space walks, and participated in the first artificial gravity experiment in space.

The polynesians knew where specific stars would rise and set. They created an accurate compass in their heads

An Indian scripture from the 5th century B.C. quotes Buddha, "Long ago ocean-going merchants ... took with them shore-sighting birds."

It is known that tame frigate birds carried messages between Pacific islands.

On a celestial sphere Greenwich Hour Angle (GHA) is the same as longitude. It is measured westward from the Greenwich meridian, from 0^0 to 360^0.

Also on a celestial sphere right, ascension is the same as longitude. It is measured from the vernal equinox, on the equator, or in the sky it becomes the First Point of Aries. It is always measured eastwards from zero to 24 hours.

If the VE is to the west of the meridian it is wrong and they have been shown on globes incorrectly for over 140 years.

The reason for this error is quite simple. Cartographers took the analemma. as it would appear in the sky and placed it on globes. This reversed the analemma and was wrong.

With coastal navigation, you use land reference points. When you draw two lines from two coastal landmarks, where these two ship to shore lines intersect is your geographic position, or fix.

Celestial navigation is similar in principle, but you are using vertical lines to the stars. With a

sextant you form a circle of equal altitude. A second circle of equal altitude will cross the first circle. A third circle will indicate a point where all three circles cross ... this is your fix.

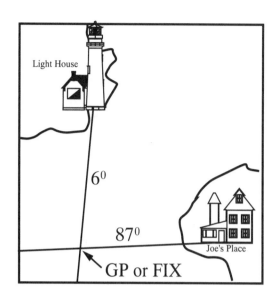

At the north and south poles, you could walk in a three foot diameter circle and cover all 24 time zones and stand in two days — but, in fact by decree, the areas are on GMT.

The author has a book in progress called: A MAGNET CALLED EARTH. Magnetism quick and easy. ISBN: 0-913257-06-0. it will show you how to calculate magnetic variation in your head, in order to correct for true north.

Before the big bang, astrophysics becomes metaphysics. We could be having a daily Genesis. It is estimated that the present world population will double by 2070. Space travel and navigation could become a necessity.

Sea turtles have been on earth over 300 million years. Adults can weigh 500 pounds and travel from Alaska to Chile. It is believed they navigate by the stars. Exactly how they do this is not known.

Why do clocks go clockwise?
The reason is because the first mechanical clock makers wanted the clock to relate to the sun dial. The shadow of the gnomon (pointer), in the northern hemisphere moves in a clockwise direction.

"Difficulties of human behavior are more difficult than the difficulties of technology." Dr. Edward Teller (father of the hydrogen bomb).

Einstein on visiting the telescope at Mt. Wilson, was told it was used to determine the structure of the universe. Mrs. Einstein said," My husband does that on an old envelope,"

Sirius is the brightest star in the heavens.

British physicist Julian Barbour believes that if nothing changes there would be no time. "Time is nothing but change... time does not exist."

On visiting Machu Pichu (little hill) in Peru, I noted a trapezoidal opening in a wall. At sunrise on the summer solstice, the sun shines through the opening and forms a shadow on a straight line, carved in a sacred stone.

Because the sun is three million miles closer during the winter solstice, in the northern hemisphere... the winter of the northern hemisphere is slightly warmer.

The oldest globe in existence was made by Martin Behaim in 1491. It doesn't show America, but it shows Japan, the equator and the tropics.

The diameter of the super-giant Betelgeuse is larger than the diameter of the orbit of Mars.

Alan Bean, "the Reminqton of space" and the fourth man to walk on the moon, asked me to check the sunlight in his studio.

Using the analemma I was of some small help in orienting his studio.

Five million pounds of material is orbiting in space... 95% is junk.

A cosmic year is 250 million earth years. As stated by the San Francisco Astronomical Society. It is the time for the sun to revolve around our galaxy.

If the sun were reduced to a two foot diameter, the earth would be the size of a pear, 215 feet away. Pluto would be the size of a pea, 1.6 miles away. The nearest star would be 10,000 miles away. Because stars are so far away, they look like points. Planets look like discs and seem to move against fixed stars.

Once you see the mountains of Mona Kea or Haleakala, you can then get out your street map.

"The Greeks of the 5th century are our contemporaries; we are no wiser than they are."
Mortimer Adler

The "line of the nodes" is simply the line between the two intersecting points between the ecliptic (orbit of earth) and the orbit of the moon. These two points are on opposite sides of the earth's orbit. The "line of the nodes" crosses on the analemma at 9* north latitude. The reason the analemma crosses at this point is because the equation of time at this point is zero. Sun time and clock time are exactly the same. The only other times when the points on the earth's eliptical orbit are exactly in sync with clock time at 12 noon, are at the two solstices. The result is that these four positions of the sun are exactly on the analemma's meridian.

In 1772, King George the fifth, had to threaten The Board of Longitude with a personal appearance if they did not give John Harrison the prize money of 20,000 pounds for finding longitude, King George was an amateur astronomer and said, "By God Harrison, I shall see you righted." John Harrison received his prize money, 43 years after he started his search.

As recently as the year 2000, astronomers thought that one point in the sky was a star. With the Hubble space telescope (HST), they discovered that the point was actually over 1,500 galaxies, representing billions of stars. Truly, new truths of very old creation.

Dr. Rita Barnabei, with the Gran Sanso National Laboratory, east of Rome, believes that WIMP (weakly interactive massive particles), or dark matter, makes up 80% of all the mass in the universe. The WIMP weighs over 50 times more than a proton.
It is believed that dark matter holds the galaxies together and bends light.

When you find water it is a leak,
When the water finds you it is a flood.
U.S. Navy definition

Actual radio conversation released by the Chief of Naval Operations on 10 Oct. 1995. Between a U.S. Navy Ship and Canadian authorities, off the coast of Newfoundland:

U.S. Ship: Please divert your course 0.5 degrees to the south to avoid collision.

CND reply: Recommend you divert your course 15 degrees to the South to avoid collision.

U.S. Ship: This is the Captain of a U.S. Navy Ship. I say again divert your course.

CND reply: No, I say again, you divert YOUR course.

U.S. Ship: THIS IS THE AIRCRAFT CARRIER U.S.S. CORAL SEA. WE ARE A LARGE WARSHIP OF THE U.S. NAVY. DIVERT YOUR COURSE NOW!!

CND reply: This is a lighthouse. Your call.

TO MEASURE A POND

From schoolboy plane geometry we know that 'in any triangle a line which is parallel to a side, divides the other two sides proportionately. These principles can be used in navigation.

IF AB IS THE UNKNOWN AND THE OTHER LINES CAN BE MEASURED THEN THE PROPORTION IS STATED:

CE IS 1/2 EB
DE IS 1/2 EA
Solve for CD
CD IS 1/2 AB

AB IS TO BD
AS
AC IS TO CE
OR

$$\frac{AB}{BD} = \frac{AC}{CE}$$

AND SOLVE FOR AB.

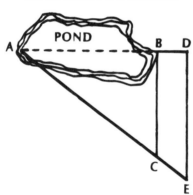

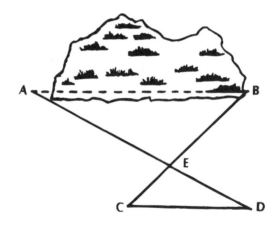

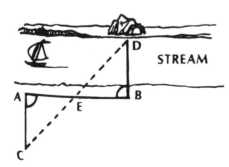

Measure a stream by creating two equilateral triangles

AE is to EB
as
AC is to BD

AE = EB
AC = BD

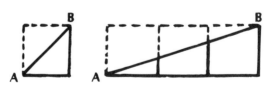

If you have either a square, or a rectangle (of land) and draw a diagonal line such as **AB**, the areas in the square and the rectangle are each divided equal.

Scotish smacks

The earth revolves around the sun at 18 miles per second, 1080 miles per minute, or 64,800 miles per hour.

All matter came from the primordial galaxy, stardust ... whether it be our galaxy, our earth, a Nobel laureate, a sextant, or a can of beer.

If you know why the earth's analemma is a figure eight and Mar's is a tear drop...you understand this book.

In talks with navigators who have sailed the oceans of the world and men who have been to the moon... they all have mentioned that their ability to think in three dimensions has helped them with celestial navigation.

Claudius Ptolemy designed the grid for earth (latitude and longitude), that we still use today.

It would take 100,000 light years to traverse our Milky Way galaxy.

"There can be no thought of finishing, for arriving at the stars...there is always the thrill of first beginning." - Dr. Robert Goddard, space pioneer.

In the 16th century, the youngest boy on board a ship, was responsible for turning the half-hour sand clock. Distance was measured as. "so many glasses away." Today a ferry trip from San Francisco to Sausalito is, "two martinis away."

We speak of *woof* and *warp*. You can think of *woof* as latitude and *warp* as longitude.

The nearest galaxy to ours is the Andromeda galaxy.

Hubble showed us that the universe is expanding. He also concluded that there is no center to the universe.

Columbus estimated 10^0 of longitude equal to 45NM, instead of the correct 60NM. Sailing from the Canary islands at west 15^0 longitude to San Salvador, at west 75^0 longitude A total voyage of about 60^0 longitude. This represents an error of 900NM, or multiplying by 1.15 equals an error of 1,035 statute miles.

Simultaneous depends on where you are.

100 nautical miles = 115 statute miles

GPS (Global Positioning Systems) or electronic navigation systems can produce errors due to atmospherics, or on board electronic disturbances.

Before the 17th century there was no need for a second of time. The second came into use because of celestial navigation.

If you understand that the sun is at the summer solstice on June 21st...and from the analemma you learn it is at 23 $1/2^0$N, on the Tropic of Cancer. You know of dead reckoning.

Longitude is simply relating time to degrees, and degrees determine longitude, and longitude degrees on a great circle, determine distance.

With the Hubble telescope you can see a dime from 100 miles away.

It takes 2 seconds for light to go from the moon to your eye and eight minutes to go from the sun to your eye.

Celestial navigation and astronomy are empirical rather than experimental. You can only observe stars you can't put them in a test tube.

Einstein's last words, "Is the universe friendly?"

Columbus noted that his compass was at 18^0 variance when he checked for true north, with the north star. If there is enough interest, a future book will explain the significance of this fact.

It is about a 3 year round trip to Mars from the moon. Should be going about 2020. Mars is about 234 miliion miles from earth.

Bees eyes can detect the polarized light of the sun, even on cloudy days and this aids in their navigation.

The space between the stars is mostly hydrogen gas.

All the planets are moving in the same direction.

If the sun was 4 inches in diameter, the nearest star would be 1,500 miles away.

Attending a talk by Dr. Edward Teller, he stated, "Low flying satellites last three years, cost one million and could be used instead of one billion dollar, high-flying satellites."

This book has tried to present neologisms, new ways of looking at things.

If the earth were the size of an apple, its atmosphere would only be as thick as the skin.

It takes about 250 million years for the sun to make one orbit around our galaxy.

The sun converts 4 million tons of matter into energy every second. It should last about 10 billion years.

Nobody knows why all the planets rotate.

Because of lower gravity, it is believed that children growing up on the moon will grow taller.

"What really interests me is did God have any choice in the universe?" Einstein

The theory of relativity predicts that an object that travels faster than the speed of light would go backwards in time.
In the words of Arthur Buller's limerick:

> *There was a young lady named Bright,*
> *whose speed was faster than light*
> *she set out one day,*
> *in a relative way,*
> *and returned home*
> *the previous night.*

Nobody knows what makes gravity happen.

"No matter, no gravity, only time, a vast sheet of space." Einstein

Mike Collins, astronaut, "I bet you don't know what ocean-earth interface is called."

Answer, "The beach"

"Buzz" Aldrin, astronaut was one of the first to suggest "*ecliptic reference*" for space flight.

After a ten year analysis of space navigation, the author has a new system involving *"point vectors."*

Every celestial body can easily be given defined points on its surface. When these points are connected from the center of the body and extended to infinity...they create line vectors.

These vectors (two numbers) from two different celestial bodies, plus the exact time can produce a GP (geographic position) in a fourth dimensional universe.

The polynesian reed and shell, stick charts, were used only by teachers and beginners.

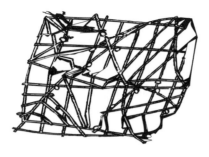

CMC: Command Module Computer

For spacecraft navigation aboard the Apollo missions, they used: 37 stars, the earth, the sun and moon. Each was given a computer number and plotted. There is no time to dwaddle when the spacecraft is moving at 3.000 feet per second.

"The view of the moon that we're having is really spectacular. It fills about three-quarters of the hatch window and of course we can see the entire circumference, even though part is in complete shadow and part is earthshine. It's a view worth the price of the trip." Neil Armstrong

Dick Gordon, pointed out in talks with him, that when going to the moon he could orient himself in space by looking back and seeing both the earth and sun at the same time. Only in outer space can you look at the sun and watch the earth rotate.

Dicks's method of orientation is not unlike the polynesian and viking navigators.

He also pointed out that on the moon, the earth is about a quarter of a million miles away ... compared with the sun at 93 million miles. The earth appears about four times bigger than the sun.

If you are at the edge of the ocean and your eye is at 6 feet, you can see the horizon at 3 miles. This is the reverse of not being able to see a 6 foot man, 3 miles away, when your eye is at sea leve. The 6 to 3 rule

The earth's angular width from the moon is about 2^0. The thickness of your finger can block out the earth.

Declination of the sun is the angular distance of the sun, north or south of the equator, along a meridian.

The chaos theory tells us life is uncertain. Eat dessert first.

'It's a strange, eerie sensation to fly a lunar landing trajectory - not difficult, but somewhat complex and unforgiving." Neil Armstrong

REFSMAT. An acronym referring to: **ref**erence **s**table member **mat**rix. As there is no sense of "up" in a spacecraft, matrix refers to angles to navigation stars.

Apolune is the highest point of a spacecraft in a lunar orbit.

PGNCS is an acronym for a spacecraft, **P**rimary **G**uidance, **N**avigation and **C**ontrol **S**ystem.

If the sun were reduced to a two foot diameter, the earth would be the size of a pear, 215 feet away. Pluto would be the size of a pea, 1.6 miles away. The nearest star would be 10,000 miles away.

As mentioned, 3 hand spans equal 90^0 from the vertical in the sky to the horizon. So it is that 6 hand spans, represent one half of a circle or 180^0. Hand spans can be a quick dead recckoning check, from north, when you do not have a compass.

The planets in our solar system are named after Greek and Roman gods.

Hand measurements are thousands of years old. Horses are measured by "hands", one hand is 4 inches.

A mile ("mille" in Latin) was originally 1,000 paces of a Roman soldier.

One nautical mile (NM) equals: 2025.4 yards, 1.852 km, 1852 meters or 6076.1 feet, and has been adopted internationally. The United States officially adopted this value in 1959.

It is interesting to note that as our knowledge of the circumference of earth increases, the nautical mile is changed and not the number of degrees in a circle.

The equitorial diameter of earth is 26 miles greater than the polar diameter.

The Apollo spacecraft took about 3 days to get to the moon. At the same speed, it would take one million years to get to the nearest star.

In distance one light year is equal to about 20,000 round trips from earth to Mars.

If there are 10,000 technological civilizations in the Galaxy...a number considered reasonable ... then one in a million Sun-like star systems should be on the air. Hardware now on the horizon is just what's necessary to score that one-in-a-million shot."

Seth Shostak
Editor SETI
News Search for Extraterrestrial Intelligence
Mt. View, CA
http://www.seti.org

"...the lifetime of the central star may become the limiting factor for the emergence of intelligence elsewhere."

Armand Delsemme
Astronomer University of Toledo
Author of Our Cosmic Origins

"Knowledge of the universe brings me no end of pleasure."

Neil de Grasse Tyson
Astrophysicist
Author of The Sky is not the Limit
Director, Hayden Planetarium
New York City, NY

In six months of daylight at the north pole, the sun travels around the sky parallel to the horizon. It is said that a botanist raised sunflowers at the north pole and they did quite well for awhile ... but since they like to face the sun...they strangled themselves to death.

Astronaut Lovell, lost power on Apollo XIII guidance systems,.."had to fly by the seat of my pants." He manually guided the spacecraft within a 2^0 window, to successfully land on earth.

In future millenniums, if we can eventually navigate to the edge of the finite universe it will present three questions:

1. *What is beyond?*
2. *What do you call it?*
3. *How far does it extend?*

Stephen Hawking has sold eight million copies of *A Brief History of Time*.

The most exciting two days in the author's life...sitting down...were spent with Dick Gordon, a man who has been to the moon, as he listened to my ramblings on celestial navigation.

In 1479 the Portugese govenment had a law that brought death by drowning to any sailor or navigator who described the secrets of their maps.

We believe the only flat map with an analemma is made by Geochron of Redwood City, CA. They produce an attractive lighted wall map, that moves to indicate the world areas of day and night.

Although parallel lines of latitude look on a globe look like simple circles. they are in effect the top circle of a cone.
There are only 2 latitudes that are not derived from cones. They are the great circle of the equator, that can go out to infinity and a straight line going to infinity between the north and south poles.

The latitude cone starts with the apex at the center of the earth. becomes a parallel of latitude at the surface of the earth and then extends to infinity.

In logic, we come back to an infinite universe, as Einstein would put it...it is more simple and more beautiful.

Dr. Edward Teller has stated, in answering a question of the author, "I do not think that God is so uninterested as to not have other beings in the universe."

If space is infinite, then with the thinking of Einstein and Teller there could be an infinite number of universes. It may be of interest to consider these universes in a Nautilus shell spiral configuration, using the Fibonacci numbers of nature. The ultimate formula for beauty and simplicity.

INTERESTING WEBSITES
FOR FURTHER STUDY

The analemma
http://www.analemma.com/

University Portsmouth, UK Cosmology
http://euclid.sms.port.ac.uk/cosmos/
cosmos.html

Dennis de Cicco's photo of analemma.
http://www.uwm.edu/People/kahl/Images/
Weather/Other/analemma.html

Sundials on the Internet.
http://www.sundials.co.uk/index.htm

Stonehenge.
http://www.sonic.net/yronwode/
stonehenge.html

Teacher's guide to navigation.
http://www.usgs.gov/education/learnweb/
MapsTGuide.html

U.S. Coast Guard Navigation Center
http://www.navcen.uscg.mil/

Celestial navigation intro.
http://peck.ipph.purdue.edu/al/Space.html

How celestial navigation works.
http://www.seatape.com/303doc.htm

Celestial navigation resources
http://riemann.usno.navy.mil/AA/faq/docs/
celnav.html

School of ocean sailing.
http://www.sailingschool.com/welcome.html

Equation of time.
http://www.faqs.org/faqs/astronomy/faq/part3/
section-15.html

Time in astronomy
http://www.faqs.org/faqs/astronomy/faq/part3/
index.html

Introduction to astronomy.
http://www.physics.emory.edu/Faculty/Ander-
son/astronomy/index.html

Ask an astronomer.
http://www.outerorbit.com/ask/ga007.htm

Wayfinders
http://www.pbs.org/wayfinders/wayfinding.html

Sky and Telescope magazine.
http://www.skypub.com/

Ocean Master's links.
http://www.oceanmasters.com/links.htm

Sailing school in Maine.
http://www.sailingschool.com/

The Navigation Foundation.
http://www.olyc.com/navigation/navfound.htm

Cruising stories.
http://www.cruisenews.net/

Newport Sailing Association.
http://www.nosa.org/

The history of time.
http://physics.nist.gov/Genlnt/Time/time.html

Ancient mariner gifts
http://amariner.com/page4.htm

Celestial navigation options.
http://www.nav.org/cel/introduction.html

Ancasta boat sales.
http://www.ancasta.co.uk/

NASA sea project.
http://seawifs.gsfc.nasa.gov/SEAWIFS.html

London sailing project.
http://www.lsp.org.uk/

The Antique Sextant.
http://www.antiquesextant.com/

The Brass Compass.
http://www.brasscompass.com/

The Starpath School
http://www.starpath.com/index.htm

Landfall navigation.
http://www.landfallnavigation.com/

GPS overview.
http://www.utexas.edu/depts/grg/gcraft/notes/gps/gps.html

Navigation seminars.
http://www.sspboatsite.com/nav/nav04.htm

Ancient sun watching.
http://www.niler.com/sighillintroduction.html.

GMT.
http://time.greenwich2000.com/

Longitude. The book.
http://time.greenwich2000.com/

Wayfinding in the Pacific.
http://tqjunior.advanced.org/3542/index.html

Poincare
http://www-groups.dcs.st-and.ac.uk/~ history/Mathematicians/Poincare.html

Conrad, Gordon-Bean. The Fantasy.
http://www.novaspace.com/AUTO/CGB.html

American Indian Astrology
http://hanksville.org/NAresources/indices/NAknowledge.html

Cosmology University of Virginia
http://astsun.astro.virginia.edu/rjh8h/foundations/contents.html

MIT Astrology
http://arcturus.mit.edu

Ion propulsion.
http://www.discovery.com/indep/newsfeatures/spacetrv/ion.htm

Nova on longitude.
http://www.pbs.org/wgbh/nova/longitude/find/

Geochron world time, with analemma.
http://www.geochron.net

Nautical Antiques. San Diego.
http://www.seajunk.com/nautical/

The Maritime Archives.
http://pc-78-120.ridac.se:8001/WWW
Nautica?Nautica.html

Mercator's world links
http://www.mercatormag.com/links.html

SSP Navigation Seminars.
http://www.sspboatsite.com/nav/nav04.htm

Columbus and dead reckoning.
http://www1.minn.net/~eithp/dr.htm

Advanced navigation course.
http://www.nlbbs.com/sysamana/ocean.htm

Celestial navigation before 1400.
http://www.ruf.rice.edu/~feegi/astro.html

A short guide to navigation.
http://home.t-online.de/home/h.umland/

Saunders & Cooke antique instruments.
http://www.saundersandcooke.com/

Lindisfarme sundials.
http://www.lindisun.demon.co.uk/

The story of sundials.
http://www.mcs.csuhayward.edu/~malek/
Mathlinks/Sundials.html

Sundial links.
http://www.ph-cip.uni-koeln.de/~roth/
slinks.html

The Mariner's museum.
http://www.mariner.org/age/menu.html

15th Century navigation.
http://www.ruf.rice.edu/~feegi/site_map.html

Equitorial coordinates.
http://www.stcloud.msus.edu/~physcrse/astr106/
declination.html

Earth and moon views.
http://www.fourmilab.ch/earthview/vplanet.html

Columbus links.
http://www.win.tue.nl/~engels/discovery/
columbus.html

Stars as gifts.
http://www.celestialregistry.com/

Celestial atlas.
http://www.stub.unibe.ch/stub/koenig/
celestial.html

Why celestial navigation?
http://celestaire.com/page4.html

Stanley London brass sextants.
http://www.stanleylondon.com/

The science of navigation.
http://www.ruf.rice.edu/~feegi/science.html

The planetarium.
http://www.mysticseaport.org/visiting/minitour/
planetarium.html

The celestial equator
http://www.stcloud.msus.edu/~physcrse/astr106/
emapequat.html

Bill Myers on navigation.
http://www.nav.org/

GMT and time keeping.
http://tycho.usno.navy.mil/time.html

Stars and constellations.
http://www.astro.wisc.edu/~dolan/constellations/
constellations.html

Positional astronomy.
http://www.sweethome.de/giesen/SME/details/
basics/index.html

Nainoa Thompson, celestial navigation without instruments.
http://leahi.kcc.hawaii.edu/org/pvs/rapanui/
nainoa.html

Biography of Kepler.
http://www-groups.dcs.st-and.ac.uk/~history/
Mathematicians/Kepler.html

Kepler's laws animated.
http://www.cvc.org/science/kepler.htm

Celestial navigation fundamentals.
http://www3.fast.co.za/~alistair/inav_c11.htm

Polynesian Voyaging Society.
http://leahi.kcc.hawaii.edu/org/pvs/

Sir Francis Drake.
http://www.mariner.org/age/drake.html

NASA and the universe.
http://www.srl.caltech.edu/seus/

Jet Propulsion Laboratory.
http://www.jpl.nasa.gov/

Celestial navigation terms.
http://www.ozemail.com.au/~jjjacq/sundry/
navcel.html

Analemmatic sundials.
http://www.tour.arizona.edu/new/museums/
museums2.shtml

World Map Links.
http://www.omnimap.com/

University of Arizona planetarium.
http://www.flandrau.org

The *Discovery,* convict ship, lying at Deptford;
The vessel that accompanied Captain Cook on his last voyage.

ORDER FORM

QUANTITY	ITEM	EACH PRICE	TOTAL
	DROPZONE PRESS · BOOKS		
	REAL ESTATE QUICK AND EASY · FULLY ILLUSTRATED 400 PAGES • 15TH EDITON ISBN: 0-913257-14-1	$24.00	$
	WIT AND WISDOM • 215 PAGES (ILLUSTRATED) OVER 2,000 OF THE BEST ONE LINERS • RETAIL $12.95	$4.95	$
	CELESTIAL NAVIGATION QUICK & EASY · FULLY ILLUSTRATED ISBN: 0-91357-11-7	$14.00	$
	DROPZONE VIDEOS · VHS		
	EGYPT · NILE CRUISE FROM CAIRO TO ABU SIMBLE · TRT: 107 MINUTES ISBN: 0-913257-2	$13.95	$
	KENYA · THE LAND · THE PEOPLE · THE ANIMALS · TRT: 60 MINUTES ISBN: 0-913257-04-4	$13.95	$
	KOH SAMUI · ISLAND OF SMILES · TRT: 14 MINUTES	$14.95	$
	SKYHOOK 5 · BALLISTIC PARACHUTES · TRT: 14 MINUTES ISBN: 0-93257-15-X	$9.00	$
	THE PARACHUTE RAID ON LOS BANOS · TRT: 21 MINUTES ISBN: 0-913257-15-3	$14.95	$
	DROPZONE POSTERS (TUBE OF 10 POSTERS · $19.00)		
	THE AMAZING ANALEMMA · 22" X 27" · 4 COLOR UNIVERSE · ECLIPTIC · ZODIAC	$10.00	$
	OTHER DROPZONE PRODUCTS		
	ANALEMMA CARD · 4" X 6" · 2 COLOR · PLASTIC · THE VIKINGS'S SECRET · A CELESTIAL NAVIGATION ANALOG (CARTON OF 50 = $25.00)	$2.00	$
		TOTAL ORDER	$

I have enclosed my check or money order. I understand each price includes shipping and handling charges.
Please, make checks payable to:
R. Maloney, DZ PRESS • Fax (415) 921 -6776

SEND ORDER TO: (Please print)

NAME_____

ADDRESS_____

E-mail_____

CITY_____

STATE_____ ZIP_____

PHONE _____ - _____
      (AREA CODE)

**DROPZONE PRESS UNCONDITIONAL GUARANTEE: YOU MUST BE SA   TISFIED, IF YOU FEEL ANY PRODUCT IS NOT SATISFACTORY,
YOU CAN RETURN IT WITHIN 30 DAYS FOR A FULL AND IMMEDIATE REFUND.**

Please, fold this envelope as indicated on other side and staple or seal.

FOLD

FROM:

NAME_____

ADDRESS_____

CITY_____

STATE_____ZIP_____-_____

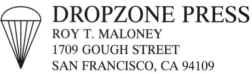

DROPZONE PRESS
ROY T. MALONEY
1709 GOUGH STREET
SAN FRANCISCO, CA 94109

FOLD

ORDER FORM

QUANTITY	ITEM	EACH PRICE	TOTAL
	DROPZONE PRESS • BOOKS		
	REAL ESTATE QUICK AND EASY • FULLY ILLUSTRATED 400 PAGES • 15TH EDITON ISBN: 0-913257-14-1	$24.00	$
	WIT AND WISDOM • 215 PAGES (ILLUSTRATED) OVER 2,000 OF THE BEST ONE LINERS • RETAIL $12.95	$4.95	$
	CELESTIAL NAVIGATION QUICK & EASY • FULLY ILLUSTRATED ISBN: 0-91357-11-7	$14.00	$
	DROPZONE VIDEOS • VHS		
	EGYPT • NILE CRUISE FROM CAIRO TO ABU SIMBLE • TRT: 107 MINUTES ISBN: 0-913257-2	$13.95	$
	KENYA • THE LAND • THE PEOPLE • THE ANIMALS • TRT: 60 MINUTES ISBN: 0-913257-04-4	$13.95	$
	KOH SAMUI • ISLAND OF SMILES • TRT: 14 MINUTES	$14.95	$
	SKYHOOK 5 • BALLISTIC PARACHUTES • TRT: 14 MINUTES ISBN: 0-93257-15-X	$9.00	$
	THE PARACHUTE RAID ON LOS BANOS • TRT: 21 MINUTES ISBN: 0-913257-15-3	$14.95	$
	DROPZONE POSTERS (TUBE OF 10 POSTERS • $19.00)		
	THE AMAZING ANALEMMA • 22" X 27" • 4 COLOR UNIVERSE • ECLIPTIC • ZODIAC	$10.00	$
	OTHER DROPZONE PRODUCTS		
	ANALEMMA CARD • 4" X 6" • 2 COLOR • PLASTIC • THE VIKINGS'S SECRET • A CELESTIAL NAVIGATION ANALOG (CARTON OF 50 = $25.00)	$2.00	$
		TOTAL ORDER	$

I have enclosed my check or money order. I understand each price includes shipping and handling charges.

Please, make checks payable to:

R. Maloney, DZ PRESS • Fax (415) 921 -6776

SEND ORDER TO: (Please print)

NAME_____

ADDRESS_____

E-mail_____

CITY_____

STATE_____ZIP_____

PHONE _____ - _____
            (AREA CODE)

DROPZONE PRESS UNCONDITIONAL GUARANTEE: YOU MUST BE SA   TSFIED, IF YOU FEEL ANY PRODUCT IS NOT SATISFACTORY, YOU CAN RETURN IT WITHIN 30 DAYS FOR A FULL AND IMMEDIATE REFUND.

Please, fold this envelope as indicated on other side and staple or seal.

FOLD

FROM:

NAME____

ADDRESS_____________________________________

CITY____

STATE______________ZIP________________-________

DROPZONE PRESS
ROY T. MALONEY
1709 GOUGH STREET
SAN FRANCISCO, CA 94109

FOLD

Mr. Analemma

A nickname the author has been called. More properly it should be Mr. Analemma #7, as there are at least six people that have done more on the subject.

Roy is a publisher and has written six books. He has met with five astronauts and worked for fifteen years on celestial navigation.

He is a former paratrooper and has developed a new system for ballistic parachutes.

Further details can be found on his two websites: http://www.themallsf.com
and http://www.dropzonepress.com.

LETTERS TO THE AUTHOR

THE ANALEMMA IN HEAVEN AND EARTH

After computing the shape of the analemma for various epochs, as published in SKY & TELE-SCOPE, I thought I had laid the venerable curve to rest. How wrong I was!

In 1979 Dennis di Cicco published his spectacular photograph of the analemma, and now H.J.P. Arnold has done the same. Moreover ... Claude Keith, Jr. and Rex Stage used di Cicco's photo and a computer to explain how the analemma relates to sunrise and sunset times at various latitudes. Their concept of analemma-rise and analemma-set is both charming and instructive. How could I have been so narrow-minded as to consider the analemma only at noon?

At least I did one thing right. At the solstices the Sun moves eastward relative to the mean sun; the fictitious object that sails around the celestial equator at a uniform rate to keep mean solar time... To my surprise Roy T. Maloney of San Francisco pointed out to me that the analemmas on most globes are reversed from mine ... This is true of the globe illustrating my own article, though I didn't notice it at the time...I would be most grateful to anyone who can point out a reason.

Bernard M. Oliver
Astronomer, Analemma expert
Co-founder of SETI
(Search for Extra-Tererrestrial Intelligence)

Editor's note: The above letter was sent and published in SKY & TELESCOPE magazine. In talks with Mr. Oliver it was explained that the reason for the analemma reversal, was that the correct analemma in the sky was taken and placed on globes, this reverses the image. This has been done incorrectly for over 140 years.

Thank you very much for calling our attention to the fact that the analemma shown on some Replogle globes drawn for us by Gustav Brueckman are incorrect. We had discontinued the use of the analemma on our globe maps and Mr. Brueckman had passed away before we realized the error ... he was considered to be the number one global cartographer in the United States, if not in the world.

W.C. Nickels
President Replogle Globes. Inc.
World's largest manufacturer of fine globes
Executive Offices Broadview, IL

THE SIMPLEST ALMANAC

From the very first time I gazed at a globe as a youngster, I remember noticing a strange shape somewhere in the Pacific Ocean ... A mis-shapen little figure eight drawn along the equator. I had no idea what it was. or what it was supposed to tell me. It wasn't until a few years ago that I learned this lopsided figure eight was called an analemma.

The analemma is an ingenious little graphic device which shows the declination of the sun for every day of the year. It displays the summer and winter solstices. plus the vernal and autumnal equinoxes. Not only those, but it manages to work in the equation of time - the difference between the apparent sun and the mean sun - as well in fact, it is this difference between the apparent sun and the mean sun that gives the analemma its characteristic lopsided figure eight shape.

...The handiest analemma I've seen...on a plastic card...is produced by Dropzone Press in San Francisco...

An entire sun almanac condensed into a simple figure eight. one diagram with almost as much sun data as an entire Nautical Almanac. Every celestial navigator should have one.

> **Tim E. Queeney**
> Marine Navigation Reporter
> Ocean Navigator magazine

...Because OCEAN NAVIGATOR is more technically oriented. we attract readers, who want to learn and know more. Your...project will be tremendously appealing to them.

> **Cameron Bright**
> Ocean Navigator magazine
> Portland, ME

Because of your contact with Bernie Oliver about the incorrect orientation of the analemma on globes.. I have spoken with Bernie and he will give you proper credit...

> **Dennis di Cicco**
> Associate editor
> Took the world's first photo of the analemma.
> SKY & TELESCOPE magazine
> Cambridge, MA

The AMAZING ANALEMMA poster "celebrates" the analemma, presenting facts about it and a brief text ... As a curiosity, it also shows the analemmas that the Sun would trace if observed from other planets in the solar system.

> ASTRONOMY magazine
> Waukesha, WI

Thanks very much for the poster...it looks very attractive and interesting. Congratulations on the nice letter about your work from Bernard Oliver. This support should be quite useful. I will show your poster to others who I think might be interested

Andrew Fraknoi
Executive Officer
Astronomical Society of the Pacific
San Francisco, CA

Thank you so much for your thoughtful gift. I am grateful for your generosity.

Bill Clinton
The White House

Thanks a million for the poster. it's beautiful.

John R. Fraser
CEO The Fountain of Youth
Planetarium & National Park
St. Augustine, FL

... the poster...is admirable in scope. content and aesthetic quality ... I would like to present a copy to each science teacher ...

Margaret Leeds
Assistant President
Beverly Hills High School
Beverly Hills. CA

Enclosed is a purchase order for a carton of your analemma cards...It is a wonderful poster ... This is the first time I have seen the analemmas for the other planets in our solar system.

Karen A. Gloria
Assistant to the Director
Van Vleck Observatory Wesleyan University
Middletown, CT

We appreciate your writing us...regarding our inflatable globe ... and letting us know the mistake ...

Elanor M. Goggin
Wings over the World Corp.
New York. NY

Have mounted your analemma poster ... and use it with groups every week...Our analemma inverted projection (using a mirror) is on our wall, is 5 1/2 feet tall using brass tacks.

James Hill
Director Rainwater Observatory
French Camp. MS

"...a must read book for all boating enthusiasts. Everything is there for beginning and experienced sailor alike. It is the prerequisite book for later navigation courses...well illustrated."

Ron Penzel
Gate 5 harbor
Sausalilto, CA

"It looks great to me... "

Mike S. Melin
Captain Boeing 747
United Air Lines
Mill Valley, CA

"...It is easy to read and understand...should be a text book requirement for all world sailors and travelers. Kudos to you...great job!!!"

Captain Bob Tucknott
Commander
Alameda County Sheriff's Air Squadron
Wing leader, Angel Flight West
San Leandro, CA

"Congratulations on a well written book...I did not find any errors. From a retired pilot who flew over Africa for 18 years. Well done my friend."

Captain Bruce L. Taylor
Oakland, CA

" 'Razzle Dazzle' was the name of Jack London's first boat and mine. I feel safer having read your book...You should always know where you are..."

"Jack London" George Rowan
Oakland Riviera, CA

"...You have the rare ability to synthesize the densest navigation information. You write in a very visual language. I find I want to know more about our global world."

Chic Ciccolini
president, Ciccolini Films
San Francisco, CA

"...fascinating!"

Tom Johnson
public speaker
San Francisco, CA

"...love the stuff."

Terry McGough
property investments
Vista, CA

"...it will take many readings to absorb... I would say more but, I'm speechless."

Capt. Waldron Vorhees
Owner of many, but not enough,
fine vessels
Ukiah, CA

"...innovatively conceptual book for my friend Nainoa Thompson, who has sailed extensively...on the Hokulea using celestial navigation. Old dogs need to learn new ttricks."

Barry Yap
Ocean sailor
Kauai, HI

"Having recently read Melville's 'Moby Dick...if Ahab had read your book...he may have bagged Moby rather than vice-versa...the world needs inventive ideas...each page is a joy to read."

Luther Elze
sailor
San Francisco, CA

" ...a stimulating read. Solving the ' so much to learn, so little time' syndrome...an ideal gift for mariners and adventurers."

Robert Von Heck
director
INVENTECH
patent advise
1-800-PAT-IDEA
San Diego, CA

" Brilliantly simple explanation by a most astute observer."

Lee Aaron Ward
architect (with Frank Lloyd Wright)
Submarine veteran
Quartermaster (assistant to navigator)
WWII on U.S.S. Sea Cat and Korea on U.S.S. Tiru
1-707-257-7508
Napa, CA

Dear Roy,

We would like to thank you again for so graciously making time this past Monday to appear on The Frank Foster Show. Your interview has prompted a great deal of comment and inquiry here and I would like you to know we intend to call on you again in the future for a longer visit. Again, thank you for your time and knowledge.

Pete Schafhausen
Producer
The Frank Foster Show
Simply Stated, Inc.
7316 Manatee Ave. W.
Bradenton, FL 34209

"Profound little book...inspiring to future navigators and scientists...Great contribution!!!"

Dan Sheridan
president
Top Achievers Institute
San Francisco, CA

"...It has a good format with attention - getting text and illustrations. The concepts have been simplified nicely... an important way to write such a complicated subject."

Carol Kiser
Park Ranger
San Francisco, CA
Maritime Museum National Historical Park
1-415-556-3002

" It was a pleasure being with you and the astronauts: Dick Gordon, Alan Bean and 'Pete' Conrad, at their 20th anniversary party, of the Apollo 12, the second landing mission to the moon. I know this put your 'feet to the fire' on celestial navigation and your continued research into the uses of the analemma....on the book...Well done!."

Gil Murray
Murray Company, international consultants
Los Altos Hills, CA

"...It is presented in a way to capture your interest from cover to cover...I am a mathemetician and physicist...really appreciate the mixing of art and science."

Will Hoskins
president
JBM Systems
Pittsburg, CA

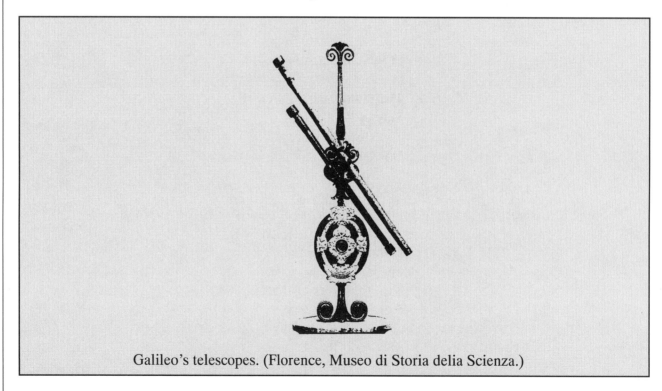

Galileo's telescopes. (Florence, Museo di Storia delia Scienza.)

NOTES ON NAVIGATION

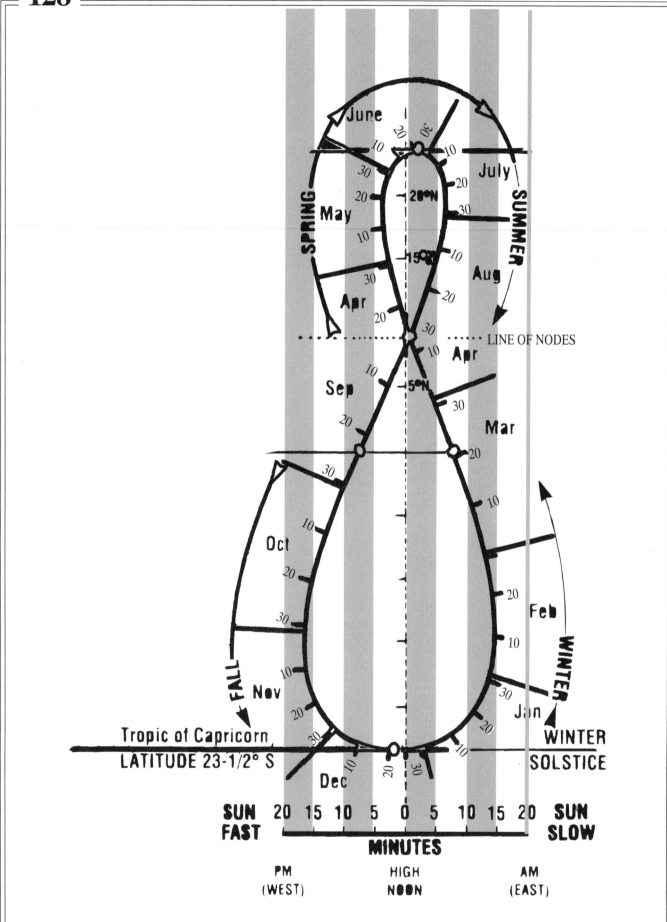